MIX
Papier aus verantwortungsvollen Quellen
Paper from responsible sources
FSC® C105338

Jawad Nasir
Muhammad Shafiq
Muhammad Mansha
Syed Hussain Haider Rizvi
Badar Ghauri

Baseline Air Quality of Azad Jammu and Kashmir

Anchor Academic
Publishing

Nasir, Jawad, Shafiq, Muhammad, Mansha, Muhammad, Rizvi, Syed Hussain Haider, Ghauri, Badar: Baseline Air Quality of Azad Jammu and Kashmir, Hamburg, Anchor Academic Publishing 2017

Buch-ISBN: 978-3-96067-150-3
PDF-eBook-ISBN: 978-3-96067-650-8
Druck/Herstellung: Anchor Academic Publishing, Hamburg, 2017

Bibliografische Information der Deutschen Nationalbibliothek:
Die Deutsche Nationalbibliothek verzeichnet diese Publikation in der Deutschen Nationalbibliografie; detaillierte bibliografische Daten sind im Internet über http://dnb.d-nb.de abrufbar.

Bibliographical Information of the German National Library:
The German National Library lists this publication in the German National Bibliography. Detailed bibliographic data can be found at: http://dnb.d-nb.de

All rights reserved. This publication may not be reproduced, stored in a retrieval system or transmitted, in any form or by any means, electronic, mechanical, photocopying, recording or otherwise, without the prior permission of the publishers.

Das Werk einschließlich aller seiner Teile ist urheberrechtlich geschützt. Jede Verwertung außerhalb der Grenzen des Urheberrechtsgesetzes ist ohne Zustimmung des Verlages unzulässig und strafbar. Dies gilt insbesondere für Vervielfältigungen, Übersetzungen, Mikroverfilmungen und die Einspeicherung und Bearbeitung in elektronischen Systemen.

Die Wiedergabe von Gebrauchsnamen, Handelsnamen, Warenbezeichnungen usw. in diesem Werk berechtigt auch ohne besondere Kennzeichnung nicht zu der Annahme, dass solche Namen im Sinne der Warenzeichen- und Markenschutz-Gesetzgebung als frei zu betrachten wären und daher von jedermann benutzt werden dürften.

Die Informationen in diesem Werk wurden mit Sorgfalt erarbeitet. Dennoch können Fehler nicht vollständig ausgeschlossen werden und die Diplomica Verlag GmbH, die Autoren oder Übersetzer übernehmen keine juristische Verantwortung oder irgendeine Haftung für evtl. verbliebene fehlerhafte Angaben und deren Folgen.

Alle Rechte vorbehalten

© Anchor Academic Publishing, Imprint der Diplomica Verlag GmbH
Hermannstal 119k, 22119 Hamburg
http://www.diplomica-verlag.de, Hamburg 2017
Printed in Germany

TABLE OF CONTENTS

CHAPTER ONE: INTRODUCTION 1
CHAPTER TWO: BACKGROUND 3
2.1 Ambient Air 3
2.2 Indoor Air Pollution 4
2.3 AJK at a Glance 5
CHAPTER THREE: METHODOLOGY 7
3.1 Data Acquisition 7
    3.1.1 Field Measurements 7
3.2 Ambient Air Monitoring Methods 9
3.3 Data Acquisition Plan 10
    3.3.1 Muzaffarabad 10
    3.3.2 Bhimber 11
    3.3.3 Mirpur 12
3.4 Site Selection Criteria 13
CHAPTER FOUR: ENVIRONMENTAL QUALITY STANDARDS 14
4.1 Description of Sources 15
    4.1.1 NAAQS/EPA 15
    4.1.2 OSHA 15
    4.1.3 WHO/Europe 15
    4.1.4 NIOSH 15
    4.1.5 ACGIH 16
CHAPTER FIVE: SAMPLING SITE DESCRIPTION 20
5.1 Muzaffarabad 20
5.2. Bhimber 24
CHAPTER SIX: MONITORING RESULTS 26
6.1 Nitrogen Dioxide (NOx) 26
    6.1.1 Sources 26
    6.1.2 Environmental and Health Effects 27
    6.1.3 Monitoring results of NO_x 27
6.2 Sulfur Dioxide (SO_2) 29
    6.2.1 Sources 29
    6.2.2 Health and Environmental Impacts of SO_2 29
    6.2.3 Monitoring results of SO2 30
6.3 Carbon Monoxide (CO) 32
    6.3.1 Health and Environmental Impacts of CO 32
    6.3.2 Monitoring results of CO 33
6.4 Carbon Dioxide (CO_2) 35
6.5 Particulate Matter (PM_{10}/$PM_{2.5}$) 38
    6.5.1 Sources 38
    6.5.2 Health effects 38
6.6 Monitoring results of PM10 and PM2.5 38
6.7 Noise and its Health effects 45
6.8 Ozone 47
6.9. Volatile Organic Compounds 50
6.10. Monitoring results of BTEX 51
6.11 Bioaerosols 57
CHAPTER SEVEN: CONCLUSIONS 61
CHAPTER EIGHT: HOURLY AVERAGE FIELD MEASUREMENT DATA 62
REFERENCES 76

CHAPTER ONE: INTRODUCTION

The Azad, Jammu and Kashmir (AJK), Figure 1, environmental profiling project has been completed aiming at the healthy environment for the citizens of AJK.

The EPA of AJK has been providing the logistical support and moblity etc., throughout the sampling period. The study "Baseline Study of Air Quality in Azad Jammu and Kashmir" has a very strong component on "Indoor Air Quality Monitoring".

The monitoring has been performed at the selected locations of Muzaffarabad, Mirpur and Bhimber (Figure 2) in residential, commercial and industrial areas for ambient air. Indoor air quality has been measured in schools, houses, hospitals and industrial units of different locations of Muzaffarabad, Mirpur and Bhimber. Geographically, northern part of AJK encompasses the lower part of the Himalayas. Fertile, green, mountainous valleys are characteristic of Azad Kashmir's geography, making it one of the most beautiful regions on the subcontinent.

The southern parts of Azad Kashmir including Bhimber, Mirpur and Kotli districts has extremely hot weather in summers and moderate cold weather in winters. It receives rains mostly in monsoon weather. Snow fall also occurs there in December and January.

This region receives rainfall in both winters and summers of the order of more than 1400 mm, with the highest average near Muzaffarabad. During summer, monsoon floods of the Jhelum and Leepa rivers are common.

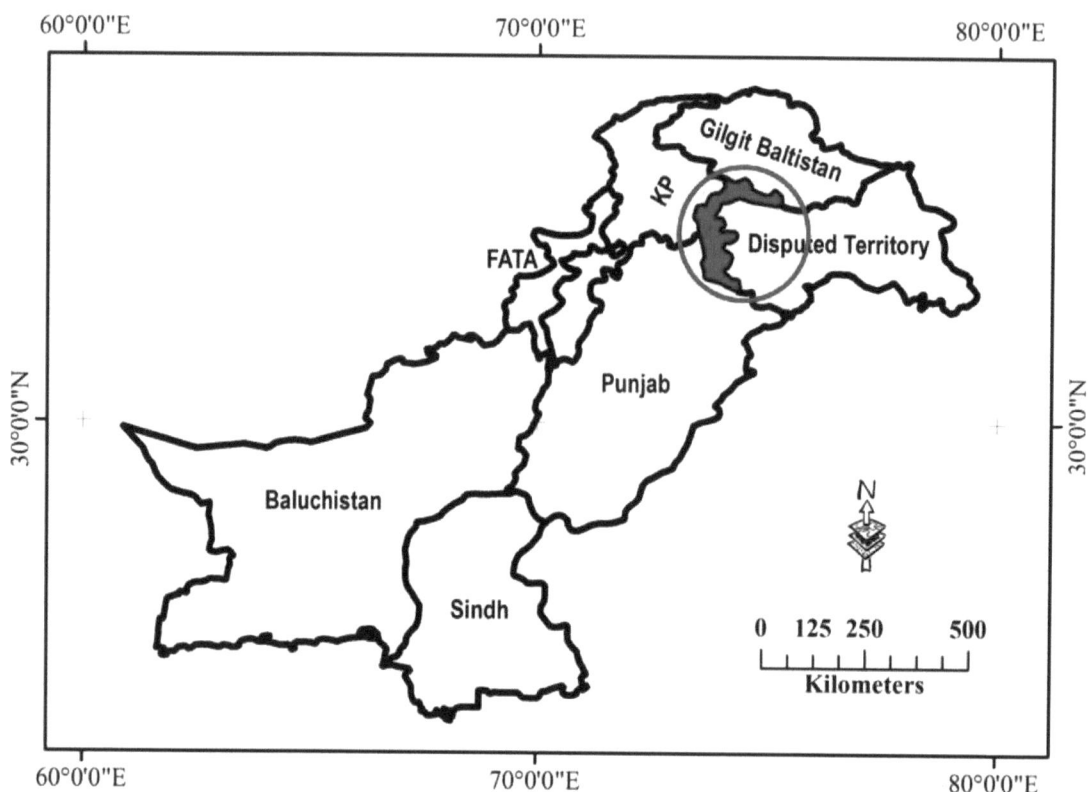

Figure 1.1: Study area

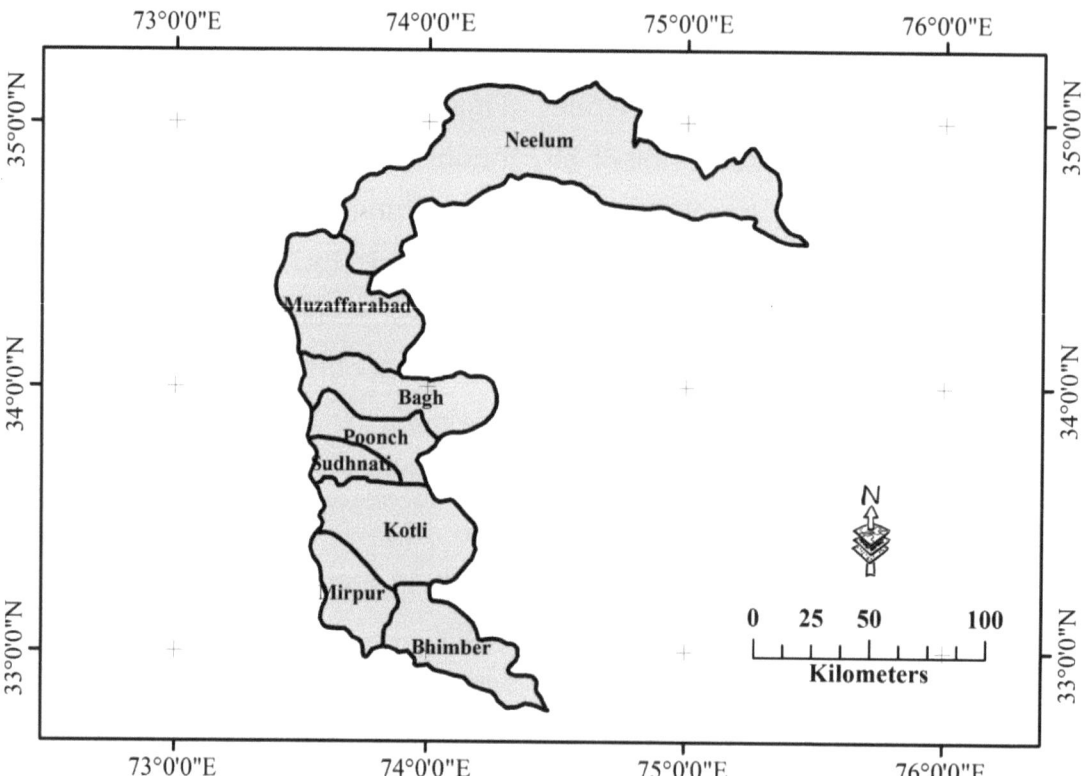

Figure 1.2: Map of districts of Azad, Jammu and Kashmir

CHAPTER TWO: BACKGROUND

Sources of air pollution cover a wide spectrum of sources from natural processes to man-made activities performed by human beings which are major contributors to air pollution. A threat to clean air is posed by the petrol and diesel fired machinery including generators, engines, etc. Similarly, transportation is another source of air pollution. Cooking and wood burring and other indoor activities are sources of indoor air pollution.

2.1 Ambient Air

Ambient air is the outdoor air in which humans and other organisms live and breathe. The contents and quality of ambient air is directly affected by the day-to-day activities of humans. Clean air is an important prerequisite for sustainable economic development and is a basic requirement for human health and welfare. Air pollution contributes to acidification and global climate change, which have an impact on crop productivity, forest growth, biodiversity, cultural monuments and many aspects of the national economy. The primary source of air pollution is the combustion of fossil fuels in vehicles, industries, power plants, and the burning of municipal solid waste.

However in AJK main sources are human activities in particular vehicular emissions, wood combustions and other domestic activities. The emissions occur from two types of sources i.e. stationary sources; and mobile sources.

The stationary sources include power plants, compressors, generators, leakages/spills of oil and gas from transmission lines and during exploration and production activities of oil and gas. Flue gases from combustion and other gas turbine processes are contaminated with harmful gases and other contaminants, which if exposed to the environment may cause deleterious effects on living beings.

Mobile sources include motorized vehicles which results in the emission of a wide variety of pollutants, principally carbon monoxide (CO), oxides of nitrogen (NO_x), oxides of Sulphur and Volatile Organic Compounds (VOCs).

Such emissions have an adverse impact on the air quality and health of human beings, particularly those who are directly exposed to this pollution.

2.2 Indoor Air Pollution

Indoor air is that which we breathe in our 'built' environment. National Health and Medical Research Council (NHMRC) of Australia defines indoor air as air within a building occupied for at least one hour by people of varying states of health. This can include the office, classroom, transport facility, shopping centre, hospitals and homes. Indoor air quality is infact defined as the totality of attributes of indoor air that affect a person's health and well being.

Relatively little attention is paid to the kinds and levels of gaseous and particulate pollutants that might be encountered in typical indoor air environment in Pakistan. Such pollutants are emitted by wood/ dung burning and kerosene stoves, dust, and fumes from paints and chemicals used in polishing industry. There is also an increasing trend toward "energy efficient" buildings that incorporate urea formaldehyde foam insulation. Carpets and decreased ventilation rates further exacerbate indoor pollution. More than 50% of households in Pakistan still use wood/biomass for cooking. It is recognized that indoor concentrations, not only of asbestos but also of the criteria pollutants such as CO, NO_2, and Particulate

Matter (PM) often exceed their urban atmosphere's outdoor levels. It is a term referring to the air quality within and around buildings and structures, especially as it relates to the health and comfort of building occupants.

IAQ can be affected by microbial contaminants (mold, bacteria). Indoor air is becoming an increasingly more concerning health hazard than outdoor air. Using ventilation to dilute contaminants, filtration, and source control are the primary methods for improving indoor air quality in most buildings. Determination of IAQ involves the collection of air samples, monitoring human exposure to pollutants, collection of samples on building surfaces and computer modeling of air flow inside buildings.

2.3 AJK at a Glance

The Azad State of Jammu and Kashmir (AJK) lies in the north of Pakistan covering an area of 5,134 square miles (13,297 square kilometers) as shown in Figure 1. According to the population census of 1998, the total population of AJK is 2.915 million, of which 88 percent is in rural and 12 percent in urban areas. The population density is 224 persons per square kilometer as against 164 in the rest of Pakistan. The next population census is due in 2017 to be conducted by the federal government.

AJK lies between longitudes 73° and 75° and between latitudes 33° and 36°. The topography is mostly hilly and mountainous with numerous valleys and stretches of plains. A large proportion of the area is under thick forest cover, with fast flowing rivers and winding streams. The forest cover is about 566,969 hectares, which is 42.6 percent of the total geographical area of AJK. The main rivers are the Jhelum, Neelum and Poonch. Elevations range from 360 meters in the south to 6,325 meters in the north. The climate is sub-tropical highland type with an average yearly rainfall of 150 cm. AJK comprises of two divisions – Muzaffarabad and Mirpur. The Muzaffarabad Division is further divided into four administrative districts (Muzaffarabad, Bagh, Poonch and Sudhnuti), whereas the Mirpur Division is divided into three districts (Mirpur, Kotli and

Bhimber) as shown in Figure 2. A total of seven industrial estates have been established in AJK, with 830 industrial units including wood works, food processing, flour mills, poultry farms, textile mills and printing presses.

CHAPTER THREE: METHODOLOGY

In order to have baseline values of different pollutants of the area this study has been conducted. Field data on ambient and indoor air pollution was collected at the selected locations of Muzaffarabad, Mirpur and Bhimber in residential, commercial and industrial areas for ambient air and in schools, houses, hospitals and industrial units for indoor air quality, by using USEPA recommended monitoring equipment.

3.1 Data Acquisition

3.1.1 Field Measurements

Field measurement data was collected for criteria pollutants Nitrogen Oxide (NO_x), Sulphur Dioxide (SO_2), Carbon Monoxide (CO), Particulate Matter (PM_{10}, $PM_{2.5}$), Ozone (O_3) as well as along with Carbon Dioxide (CO_2), VOCs [including Benzene, Toluene, Ethyl Benzene, Xylene (BTEX)], meteorological parameters, Formaldehyde, Fungal Bioaerosols, and noise levels in residential, commercial and industrial areas for ambient air of Muzaffarabad, Mirpur and Bhimber and in schools, houses, hospitals and industrial units for indoor air quality.

The measurements were above the following detection limits:

Equipment	Min. Concentration Limits
SO_2 Analyzer	~1 µg/m³
CO Analyzer	< 0.5 µg/m³
NO_x Analyzer	~ 1 µg/m³
PM_{10}/ $PM_{2.5}$ Sampler	~ 5 µg/m³
Noise Sampler	~20 dB
CO_2 Analyzer	~1 µg/m³
O_3 Analyzer	~1 µg/m³

Table: 3.1: Concentration limits of different equipment

Duration of measurements for each pollutant (CO, CO_2, NO_x, SO_2, and $PM_{10}/PM_{2.5}$) was 6 to 24 hours at each location. USEPA methods/procedures for monitoring the air quality were used.

Units of measurement for air pollutants were µg/m³ for PM_{10} while ppm & ppb for gaseous pollutants. The following conversion were used to convert ppb and ppm to µg/m³

X ppm = (Y mg/m³)*(24.45)/molecular weight
NOx Molecular Weight = 46.01(1 ppm= 1.88 mg/m³)
SO_2 Molecular Weight = 64.06 (1 ppm = 2.62 mg/m³)
CO Molecular Weight = 28.0101(1 ppm = 1.15 mg/m³)
CO_2 Molecular Weight = 44.0095(1 ppm = 1.8 mg/m³)
Y= Observed value in mg/m³

3.2 Ambient Air Monitoring Methods

Pollutants	Methods	Principle of Operations
NO_x	Reference Method in Appendix F of 40 CFR Part 50	Chemiluminescence
SO_2	Ambient Monitoring Reference & Equivalent Method of 40CFR Part 52	Fluorescence Method
CO	Method in Appendix C of 40 CFR Part 50	IR Gas Filter Correlation
CO_2	Method in Appendix C of 40 CFR Part 50	IR Gas Filter Correlation
$PM_{10}/PM_{2.5}$	Reference Method in Appendix J of 40 CFR Part 50	Gravimetric Method
VOC	Reference Method in Appendix B of 40 CFR Part 52	Gas Chromatography
O_3	Reference Method NO.EQOA-0992-087 of 40 CFR Part 53	Non dispersive UV absorption method
Noise level	Ambient Monitoring Reference & Equivalent Method 40CFR part 205	Preamplifier detector with the help of microphone

Table: 3.2: Ambient Air Monitoring Methods

3.3 Data Acquisition Plan

The work plan followed for data acquisition was as under:

3.3.1 Muzaffarabad

Site No -I: Old Secretariat

Latitude:	34°21'37.50"N
Longitude:	73°28'23.10"E
Starting Date:	25-03-2010
Starting Time:	12:00 hrs
Completion Date:	26-03-2010
Completion Time:	07:00 hrs
Sampling Duration:	19 hrs

Site No –II: Madina market

Latitude:	34°22'18.65"N
Longitude:	73°28'8.84"E
Starting Date:	26-03-2010
Starting Time:	11:00 hrs
Completion Date:	27-03-2010
Completion Time:	10:00 hrs
Sampling Duration:	21 hrs

Site No –III: AIMS Hospital

Latitude:	34°19'55.83"N
Longitude:	73°28'4.80"E
Starting Date:	27-03-2010
Starting Time:	13:00 hrs
Completion Date:	27-03-2010
Completion Time:	20:00 hrs
Sampling Duration:	08 hrs

Site No –IV: Govt Girls High School (Sehli Sarkar)

Latitude: 34°21'27.80"N
Longitude: 73°28'28.20"E

Starting Date: 29-03-2010
Starting Time: 08:00 hrs
Completion Date: 29-03-2010
Completion Time: 17:00 hrs
Sampling Duration: 9 hrs

Site No –V: Household Mujhoi (Garhi Dopatta)

Latitude: 34°15'14.9"N
Longitude: 73°35'16.8"E

Starting Date: 30-03-2010
Starting Time: 11:00 hrs
Completion Date: 30-03-2010
Completion Time: 18:00 hrs
Sampling Duration: 7 hrs

Site No –VI: Govt. Post Graduate Collage (Garhi Dopatta)

Latitude: 34°13'27.6"N
Longitude: 73°37'01.4"E

Starting Date: 30-03-2010
Starting Time: 20:00 hrs
Completion Date: 31-03-2010
Completion Time: 15:00 hrs
Sampling Duration: 19 hrs

3.3.2 Bhimber

Site No -I: Gurah Lailian/Pindi

Latitude: 32°59'02.3"N
Longitude: 74°61'00.1"E

Starting Date: 01-04-2010
Starting Time: 15:00 hrs
Completion Date: 02-04-2010
Completion Time: 13:00 hrs
Sampling Duration: 22 hrs

Site No –II: Govt. Dispensary

Latitude:	32°58'28.5"N
Longitude:	74°04'41.6"E
Starting Date:	02-04-2010
Starting Time:	14:00 hrs
Completion Date:	02-04-2010
Completion Time:	23:00 hrs
Sampling Duration:	9 hrs

3.3.3 Mirpur

Site No –I: Zahoor Food Industry (Old Industrial State)

Latitude:	33°09'03.7"N
Longitude:	73°43'15.2"E
Starting Date:	03-04-2010
Starting Time:	12:00 hrs
Completion Date:	03-04-2010
Completion Time:	22:00 hrs
Sampling Duration:	9 hrs

Site No –II: Alkhair Molti foam (New Industrial State)

Latitude:	33°06'41.10"N
Longitude:	73°43'26.80"E
Starting Date:	05-04-2010
Starting Time:	11:00 hrs
Completion Date:	06-04-2010
Completion Time:	11:00 hrs
Sampling Duration:	24 hrs

Site No –III: Nangi Adda

Latitude:	33°09'08.3"N
Longitude:	73°44'21.5"E
Starting Date:	06-04-2010
Starting Time:	12:00 hrs
Completion Date:	07-04-2010
Completion Time:	10:00 hrs

Sampling Duration: 22 hrs

Site No –VI: Household Sector F-2 (Kharak)

Latitude: 33°08'42.5"N
Longitude: 73°44'04.0"E
Starting Date: 07-04-2010
Starting Time: 10:00 hrs
Completion Date: 07-04-2010
Completion Time: 19:00 hrs
Sampling Duration: 09 hrs

Site No – V: Chaksawari Bridge

Latitude: 33°12'21.7"N
Longitude: 73°50'23.6"E
Starting Date: 08-04-2010
Starting Time: 11:00 hrs
Completion Date: 09-04-2010
Completion Time: 08:00 hrs
Sampling Duration: 21 hrs

Site No –VI: Chaok Shaheedan

Latitude: 33°08'55.8"N
Longitude: 73°44'56.3"E
Starting Date: 09-04-2010
Starting Time: 11:00 hrs
Completion Date: 10-04-2010
Completion Time: 11:00 hrs
Sampling Duration: 24 hrs

3.4 Site Selection Criteria

Monitoring sites were provided by EPA-AJK in consultation with SUPARCO and the following consideration was made for site selection

- Site is representative of major contributors of air pollutants
- Reprehensive area of the city, rural or industrial activities.
- Projected area for future proposed development.

CHAPTER FOUR: ENVIRONMENTAL QUALITY STANDARDS

This section summarizes standards and guidelines for a number of contaminants commonly found indoors and outdoors, which can be used as acceptable indoor/outdoor air quality levels. Criteria contaminants are detailed, including carbon dioxide, carbon monoxide, ozone, and particulates.

Formaldehyde, the most well-known volatile organic compound, is also included, but recommended concentrations for other volatile organic compounds (VOCs) are summarized separately recommended concentrations are provided from different agencies.

In most cases, the primary objective in setting recommended limits was to minimize health risks to the general public, or to sectors of the public, such as industrial workers or sensitive individuals. It is important to note that lower limits might be needed to avoid occupant dissatisfaction, discomfort, unacceptable odors, and sensory irritation. It is also impractical to assume that maintaining contaminant concentrations below these recommended levels will guarantee the absence of all adverse health effects for all occupants.

As the standards and guidelines given in Table differ in terms of the criteria used to set limits the population focused on, and the context for application, readers are strongly advised to consult the source documents before applying these recommendations.

4.1 Description of Sources

The standards and guidelines featured in Table are described below.

4.1.1 NAAQS/EPA

The National Ambient Air Quality Standards (NAAQS) were developed by the U.S. Environmental Protection Agency (EPA) under the Clean Air Act (last amended in 1990). These enforceable standards were developed for outdoor air quality, but they are also applicable for indoor air contaminant levels. The concentrations are set conservatively in order to protect the most sensitive individuals, such children, the elderly, and those with asthma. By law, these regulatory values must be reviewed every five years

4.1.2 OSHA

The U.S. Occupational Health and Safety Administration (OSHA) developed enforceable maximum exposures for industrial environments. The standards were developed through a formal rule-making process, and the permissible limits can only be changed by reopening this process. The Permissible Exposure Limits (PELs) given in Table are designed to protect the average industrial worker, but do not take into account the possible reactions of sensitive individuals (ASHRAE, 2004; OSHA, 2005).

4.1.3 WHO/Europe

The World Health Organization's (WHO) Office for Europe, based in Denmark, developed guidelines to be used in non-industrial settings. These guidelines were developed in 1987 and updated in 1999. They are intended for application to both indoor and outdoor exposures, but are guidelines rather than an enforceable standard (ASHRAE, 2004; WHO, 2000).

4.1.4 NIOSH

Recommended maximum exposures for industrial environments have also been developed by the U.S. National Institute for Occupational Safety and Health (NIOSH). These guidelines are published in a set of criteria documents, which contain a review of relevant literature and Recommended Exposure Limits (RELs). These non-enforceable recommendations are not reviewed regularly, and in some cases levels are set above those needed for health reasons because commonly available industrial hygiene practices do not reliably detect substances at lower levels (ASHRAE, 2004, NIOSH, 2005).

4.1.5 ACGIH

The American Council of Governmental Industrial Hygienists recommends Threshold Limit Values (TLVs) as maximum exposures for industrial environments. The TLVs are set by CMEIAQ-II Report 5.1 committee, who review the existing scientific literature and recommend guideline concentrations. The recommendations are applicable for normal industrial working conditions (i.e. 40 hours a week), and for single contaminant exposure. These recommendations are guidelines, rather than enforceable standards, and are not selected to protect the most sensitive workers.

Figure 4.1: Sampling equipment

Table 4.1: Standards and Guidelines for Common Indoor Contaminants

Parameters	NAAQS/EPA (2000)[a]	OSHA[a]	WHO/Europe (2000)[a]	NIOSH (1992)[a]	ACGIH (2001)[a]	Hong Kong (2003)[i]
Carbon dioxide	———	5,000	———	5,000 30,000 [15 min]	5,000 30,000 [15 min]	800/1,000 [8 hr]
Carbon monoxide	9 35[1 hr]	50	90 [15 min], 50 [30m] 25 [1 hr], 10 [8 hr]	35 200 [C]	25	1.7 / 8.7 [8 hr]
Formaldehyde	(See note e)	0.75 2 [15 min]	0.081 (0.1mg/m³) [30 min]	0.016 0.1 [15min]	0.3 [C]	0.024 / 0.081 [8hr]
Nitrogen dioxide	0.05 [1 yr]	5 [C]	0.1 [1 hr] 0.004 [1 yr]	1.0 [15 min]	35 [15 min]	0.021/0.08 [8 hr]
Ozone	0.12 [1 hr] 0.08	0.1	0.064 (120µg/m³) [8 hr]	0.1[C]	0.05-heavy work, 0.08-moderate work, 0.1-light work, 0.2- any work (2 hr)	0.025 / 0.061 [8 hr]
Particles <2.5µm MMAD	15 µg/m³ [1 yr] 65 µg/m³[24hr]]	5 mg/m³	———	———	3 mg/m³	———
Particles <10 µm MMAD	50 µg/m³ [1 yr] 150 µg/m³ [24hr]	———	———	———	10 mg/m³	0.02/0.18 mg/m³ [8hr]
Sulfur dioxide	0.03 [1 yr] 0.14 [24hr]	5	0.048 [24 hr] 0.012 [1 yr]	2 5 [15 min]	2 5 [15 min]	———
Total Particles		15 µg/m³				

Unless otherwise specified, values are given in parts per million (ppm)

Number in brackets [ ] refers to either a ceiling or to averaging times of less than or greater than eight hours

(min=minutes; hr=hours; yr=year; C=ceiling; L=long term.

Where no time is specified, the averaging time is eight hours.

The following table gives standards of ambient air by USEPA, WHO, World Bank and NEQS approved 1999 Standards. Proposed Ambient standards are also included in this report.

Table: 4.2 Ambient Air quality Standards (NEQS, World Bank and WHO)

Pollutants	USEPA		WHO		Pak EPA (Effective from 1st Jan 2009)	
	Averaging Time	Standard	Averaging Time	Standard	Averaging Time	Standard
SO_2	24 HRS	0.14 ppm	24 HRS	20 µg/m³	24 HRS	120 µg/m³
CO	8 HRS / 1 HR	10 mg/m³ (9 ppm) / 40 mg/m³ (35 ppm)	-	-	08 – HRS	05 mg/m³
NO_2	1 HR / ANNUAL MEAN	100 ppb (53 ppb)	1 HR / ANNUAL MEAN	200 µg/m³ / 40 µg/m³	24 HRS	80 µg/m³
NO	-	-	-	-	24 HRS	40 µg/m³
PM_{10}	24 HRS	150 µg/m³	24 HRS	50 µg/m³	24 HRS	250 µg/m³
$PM_{2.5}$	24 HRS	35 µg/m³	24 HRS	25 µg/m³	24 HRS	40 µg/m³
Ozone	08 HRS	0.075 ppm	08 HRS	100 µg/m³	01 HR	180 µg/m³

Table 4.3: Proposed Environmental Quality Standards for Noise

Standards	S. NO	Category /Area/Zone	Effective from 1st January 2009	
			Day Time Limits (dB)	Night Time Limits (dB)
PAK EPA	1	Residential	65	50
	2	Commercial Area	70	60
	3	Industrial Area	80	75
USEPA	1	Indoor	45	
	2	Outdoor	55	

CHAPTER FIVE: SAMPLING SITE DESCRIPTION

Here the description of two districts (where air sampling is performed) is provided.

5.1 Muzaffarabad

Muzaffarabad district is bounded by North-West Frontier Province in the west, by the Kupwara and Baramulla districts of on the Indian side of the Line of Control in the east, and the Neelum District of Azad Kashmir in the north. The population of the district, according to the 1998 Census, was 725,000, and according to a 1999 projection, the population had risen to almost 741,000. The district comprises three tehsils, and the city of Muzaffarabad serves as the capital of Azad Kashmir.

Muzaffarabad is situated at the confluence of the Jhelum and Neelum rivers. The city is 138 kilometres from Rawalpindi and Islamabad and about 76 kilometres from Abbotabad. Cradled by lofty mountains, Muzaffarabad reflects a blend of various cultures and languages. The main language is a form of Hindko. The Neelum River plays a dominant role in the microclimate of Muzaffarabad which joins Jehlum River near Domail.

Fig 5.1: Map of District Muzaffarabad

Fig 5.2: Windrose Model for Muzaffarabad Area during the sampling days

A weather condition at Muzaffarabad remains cloudy during the monitoring. Mostly southeasterly winds were observed during the measurements. Rainfall while monitoring was performed in **Garhi Dopatta** and Madina Market Muzaffarabad. Mirpur

Mirpur district comprises of partly plain and partly hilly areas. Its hot climate and other geographical conditions closely resemble those of Jhelum and Gujarat, the adjoining districts of Pakistan. Mirpur city is situated at 459m above sea-level and is linked with the main Peshawar-Lahore Grand Trunk road at Dina. The building of the new city in late sixties paved the way for new Mirpur situated on the banks of Mangla Lake. The city is well planned and the buildings are mostly of modern design. Mirpur is developing into an industrial city very rapidly. Textile, vegetable ghee, logging and sawmills, soap, cosmetics, marble, ready-made garments, matches, rosin, turpentine and Vespa scooter industrial units have been established in the area.

Fig No 5.3: Map of District Mirpur

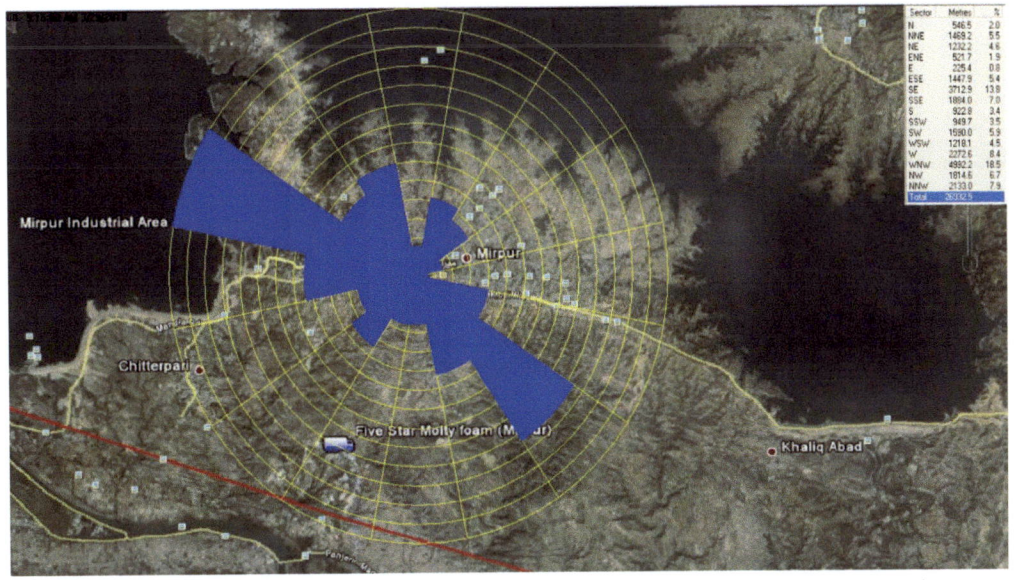

Fig 5.4: Windrose Model for Mirpur Area during the sampling days

While weather remained dry during the monitoring in Mirpur. Normal winds were prevailing during the monitoring period.

5.2. Bhimber

Bhimber is the chief town of Bhimber District, Azad Kashmir. The town is located at 32°58'60N and 74°04'0E and is situated on the border with Azad Kashmir and Pakistan at a distance of about 50 km from Mirpur, Azad Kashmir and about 166 km from Rawalpindi, Pakistan. Bhimber is also known as Bab-e-Kashmir (Door to Kashmir), due to its geographical location of the city. The area is very rich in archaeological remains due to its strategic location; it lies on the route that was followed by the Mughal Emperors for their frequent visits to the Kashmir Valley.

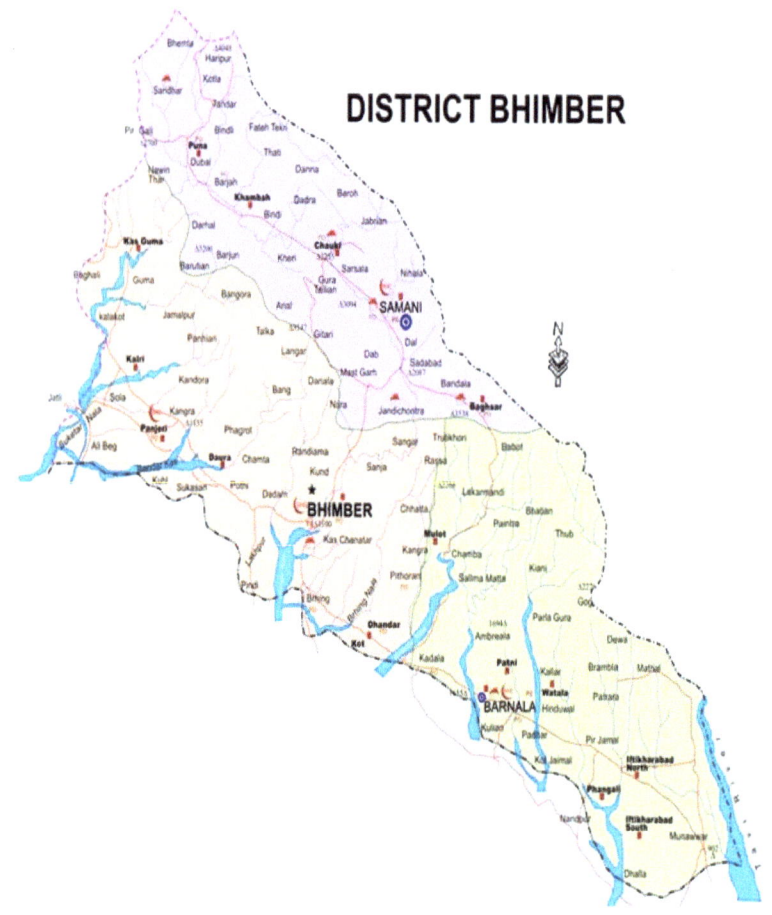

Fig No 5.5: Map of District Bhimber

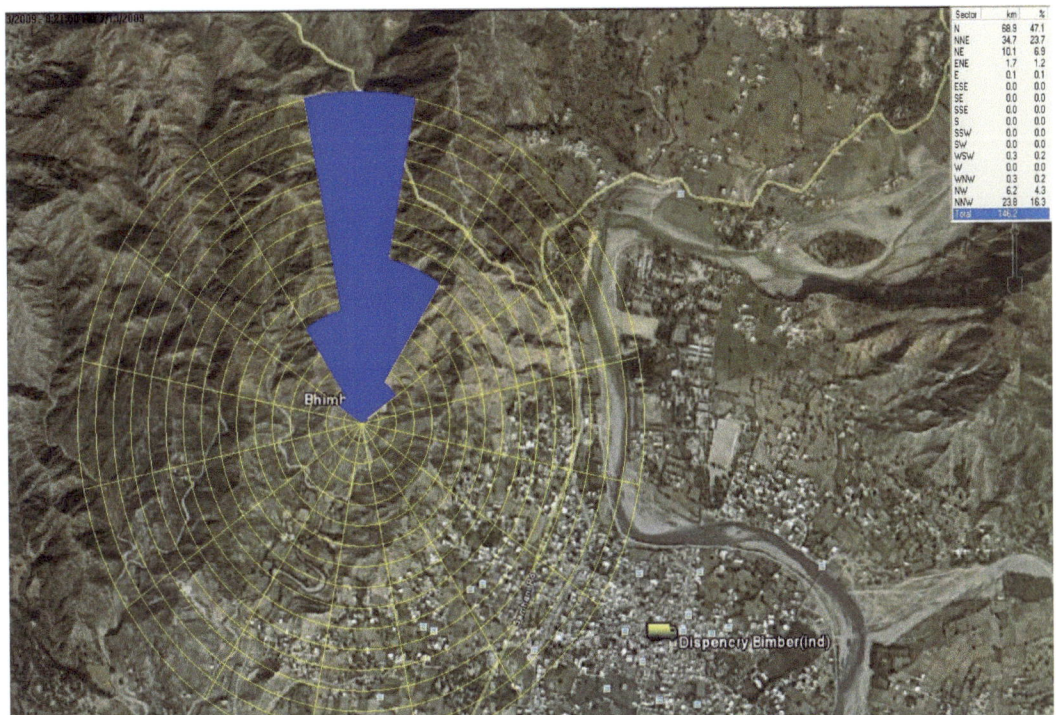

Fig 5.6: Windrose Model for Bhimber Area during the Sampling days

While weather remained dry during the monitoring in Mirpur. Normal winds were prevailing during the monitoring period.

CHAPTER SIX: MONITORING RESULTS

In this section the air pollutants like SO_2, NO_x, CO, CO_2, O_3, PM_{10}/$PM_{2.5}$, VOCs and Bioaerosols are briefly discussed which are measured at different selected locations of AJK. Hourly measured data has provided as annexure while the minimum, maximum and average values at every site is provided in the tables in the text. Data of indoor and ambient air is also provided as bar charts.

6.1 Nitrogen Dioxide (NOx)

Nitrogen oxides (NO_x) are mixture of nitric oxide (NO) and nitrogen dioxide (NO_2). Nitric oxide is colorless, odorless and is oxidized in the atmosphere to form nitrogen dioxide.

Nitrogen dioxide is an odorous brown, acidic, highly-corrosive gas that can affect our health and environment. NO_x is a critical component of photochemical smog.

6.1.1 Sources

All combustion processes in air produce oxides of nitrogen (NO_x). Motor vehicles account for about 70% of total NO_x emissions.

Indoor domestic appliances (gas stoves, gas or wood heaters) can also be significant sources of nitrogen oxides, particularly in areas that are poorly ventilated.

6.1.2 Environmental and Health Effects

Nitric oxide does not significantly affect human health. On the other hand, elevated levels of nitrogen dioxide cause damage to the mechanisms that protect the human respiratory tract and can increase susceptibility to respiratory infections. Long-term exposure to high levels of nitrogen dioxide can cause chronic lung disease. Nitrogen dioxide is harmful to vegetation, can fade and discolor fabrics, reduce visibility, react with surfaces and furnishings.

6.1.3 Monitoring results of NO_x

NO_x Concentration ($\mu g.m^{-3}$) in Muzaffarabad

S. No	Site Name	Min	Max	Average
1	Old Secretariat	40.1	86.7	62.3
2	Madina Market	28.8	58.7	43.2
3	AIMS Hospital	43.5	48.8	46.1
4	Govt Girls High School (Sehli Sarkar)	28.6	49.3	40.5
5	Mujhoi (Household, Garhi Dopatta)	60.0	105.8	78.1
6	Govt Post Graduate College	46.9	63.5	51.1

NO_x Concentration ($\mu g.m^{-3}$) in Bhimber

S. No	Site Name	Min	Max	Average
1	Gurah Lailian/Pindi	58.7	71.6	64.6
2	Govt. Dispensary	38.7	45.7	40.5

NO_x Concentration ($\mu g.m^{-3}$) in Mirpur

S. No	Site Name	Min	Max	Average
1	Zahoor Food Industry (Old Industrial State)	45.1	65.6	51.4
2	Alkhair Molti foam (Industrial State)	28.0	49.3	40.2
3	Nangi Adda	45.3	66.6	57.3
4	Household Sector F2 Kharak	44.6	65.0	53.0
5	Chaksawari Bridge	32.3	53.8	43.0
6	Chaok Shaheedan	40.0	68.6	57.5

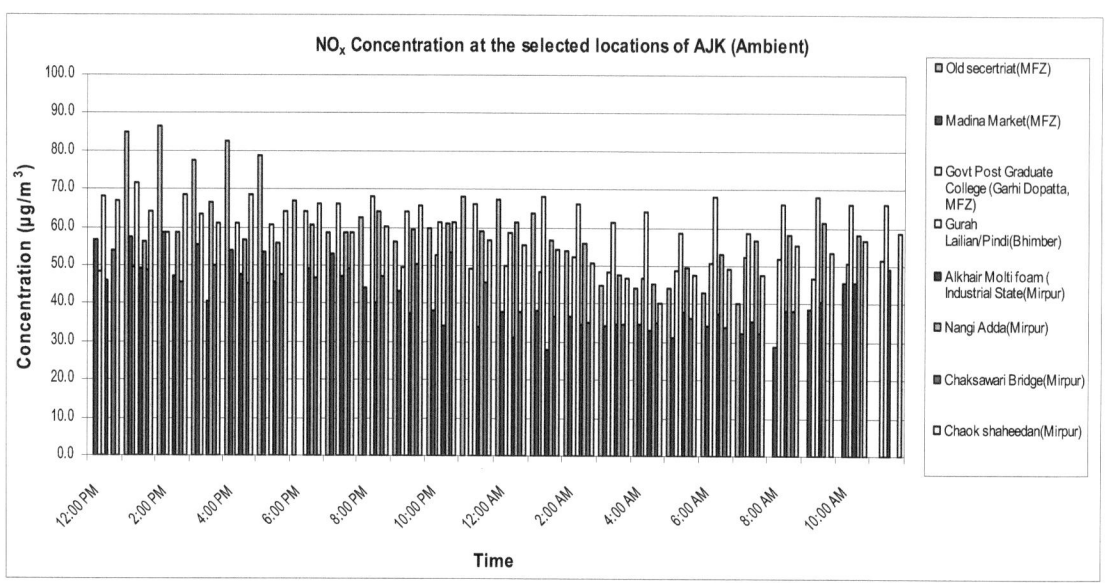

Fig 6.1

In Muzaffarabad NOx measured at various sites varied from 28.6 to 105.8 µg.m^{-3} and the average values recorded at different sites varied from 40.5 to 78.1 µg.m^{-3}. In Bhimber NOx Varied from 38.7 to 71.6 µg.m^{-3} while the average values varied from 40.5 to 64.6 µg.m^{-3} respectively. Similarly in Mirpur it varied from 28 to 68.6 µg.m^{-3} while the average values varied from 40.2 to 57.5 µg.m^{-3}. All the values recorded at different location in AJK are within the US EPA, WHO (200 µg/m^3) and NEQS (80+40 µg/m^3) values. Yousufzai et al. (2000) performed continuous measurements at the Sindh Industrial Trading Estate, Karachi and quoted values for NO and NOX in the range of 13.3–131.4 and 32.3–35.9 µg/m^3, respectively. The data included in a 1992 report (WHO/UNEP 1992) revealed that the daily average concentration of NO_2 in Karachi was 38–544 µg/m^3 respectively.

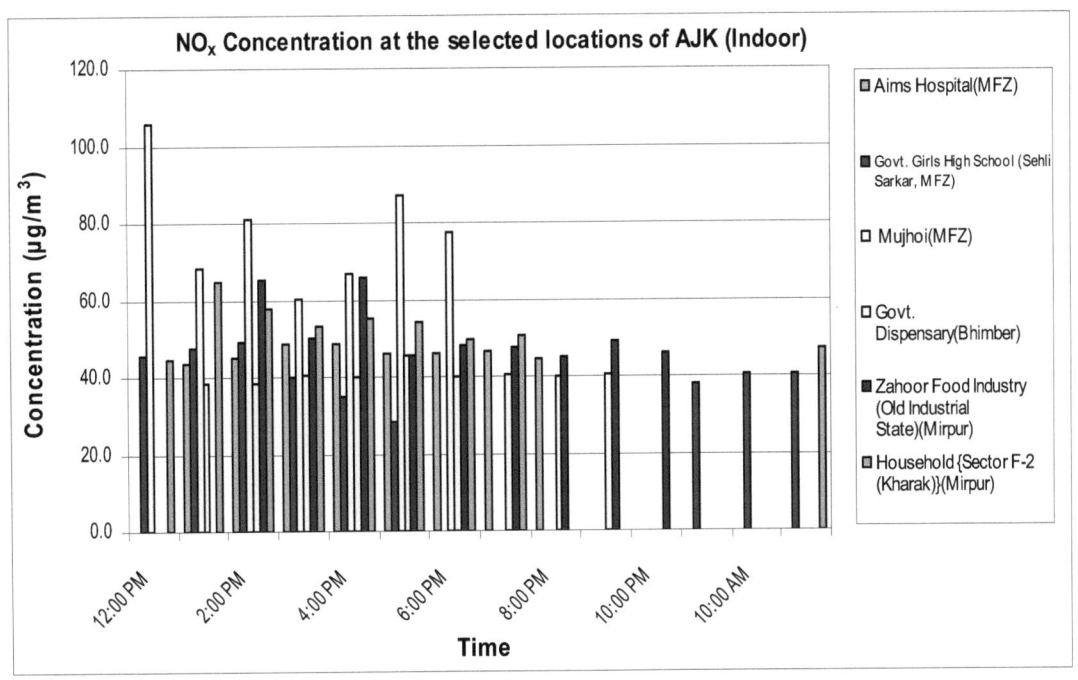

Fig 6.2

6.2 Sulfur Dioxide (SO$_2$)

Sulfur Dioxide (SO$_2$) is a colorless, extremely irritating gas or liquid, which is used in many industrial processes, especially in the manufacture of sulfuric acid.

6.2.1 Sources

Motor vehicles contribute about 80-90% of sulfur dioxide. Other sources include fossil fuel combustion, particularly coal burning power plants and industrial processes such as wood pulping, paper manufacture, petroleum, metal refining, metal smelting particularly from ores containing sulfide.

6.2.2 Health and Environmental Impacts of SO$_2$

SO$_2$ causes a wide variety of health and environmental impacts because of the ways it reacts with other substances in the air. Particularly sensitive groups include people with asthma who are active outdoors, children, the elderly and people with heart or lung diseases. Longer-term exposures to high levels of SO$_2$ gas and particles cause respiratory illness and aggravate existing heart diseases. SO_2 reacts with other chemicals in the air to form tiny sulfate particles. When these are breathed they gather in the lungs and are associated with increased respiratory symptoms and disease causing difficulty in breathing and premature death.

6.2.3 Monitoring results of SO2

SO_2 Concentration ($\mu g.m^{-3}$) in Muzaffarabad

S.No	Site Name	Min	Max	Average
1	Old Secretariat	42.4	95.2	65.9
2	Madina Market	27.0	61.6	47.2
3	AIMS Hospital	36.0	43.8	39.1
4	Govt Girls High School (Sehli Sarkar)	31.7	51.1	39.0
5	Mujhoi (Household, Garhi Dopatta)	31.4	42.4	38.1
6	Govt Post Graduate College	31.4	62.4	40.8

SO_2 Concentration ($\mu g.m^{-3}$) in Bhimber

S.No	Site Name	Min	Max	Average
1	Gurah Lailian/Pindi	59.2	74.9	63.3
2	Govt. Dispensary	25.7	31.7	28.0

SO_2 Concentration ($\mu g.m^{-3}$) in Mirpur

S.No	Site Name	Min	Max	Average
1	Zahoor Food Industry (Old Industrial State)	24.4	31.2	28.0
2	Alkhair Molti foam (Industrial State)	22.8	35.6	28.1
3	Nangi Adda	37.5	66.5	55.0
4	Household	23.3	35.4	29.8
5	Chaksawari Bridge)	29.3	51.1	40.6
6	Choak shaheedan	46.1	68.9	58.1

SO_2 measured at the different sites of Muzaffarabad varied from 27.0 to 95.2 $\mu g.m^{-3}$ and the average values recorded at different sites varied from 38.1 to 65.9 $\mu g.m^{-3}$. All the values recorded at different location of Muzaffarabad area within the US EPA, World Bank and NEQS values of 120 $\mu g/m^3$. In Bhimber the hourly values varied from 25.7 to 74.9 $\mu g.m^{-3}$ $\mu g.m^{-3}$ while the site averages are 28.0 and 63.3 $\mu g.m^{-3}$ respectively. While in Mirpur it 22.8 to 68.9$\mu g.m^{-3}$ while the average values varied from 28.0

to 58.1µg.m⁻³. Average values of all the sites are measured in AJK are within the US EPA, World Bank and NEQS values of 80µg/m³ (annual) and 120µg.m⁻³(24hrs) respectively.

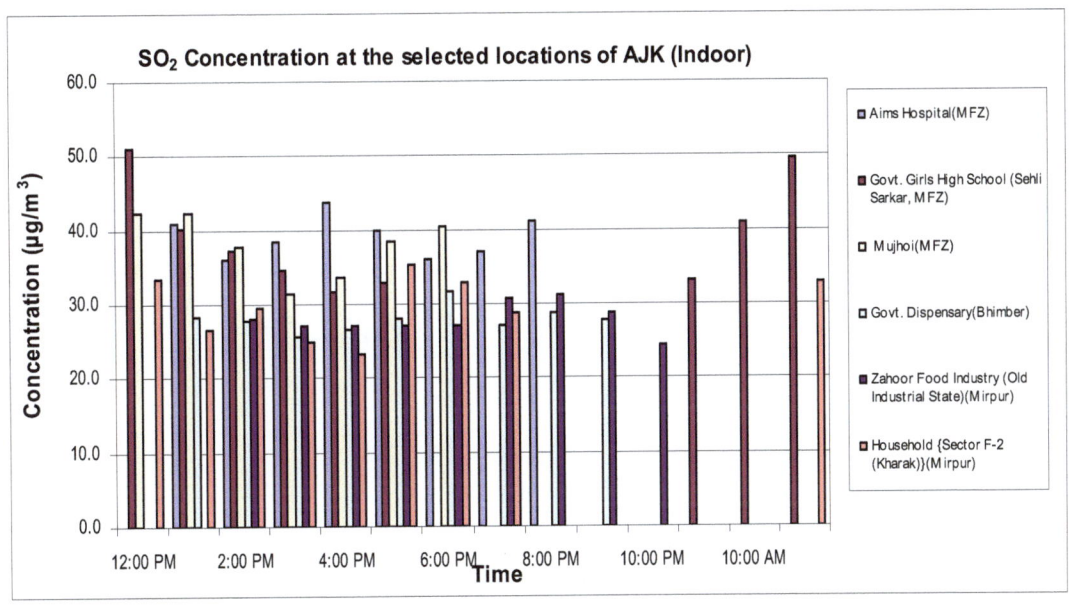

Fig 6.3

GEMS measurements from Lahore showed that the annual SO_2 concentration, in the city center, for 1978 was 49µg/m³ while 40µg/m³ was recorded at a suburban residential area during 1979 (WHO 1984). Hashmi and Khan (2003) reported that levels of SO_2 at Korangi Industrial Area and Sindh Industrial Trading Estate (Karachi) were 7.4 and 4.9µg/m³, respectively. At Port Qasim (Karachi) for 7 days during November, the concentration of SO_2 was 6.3µg/m³ (Hashmi et al. 2005).

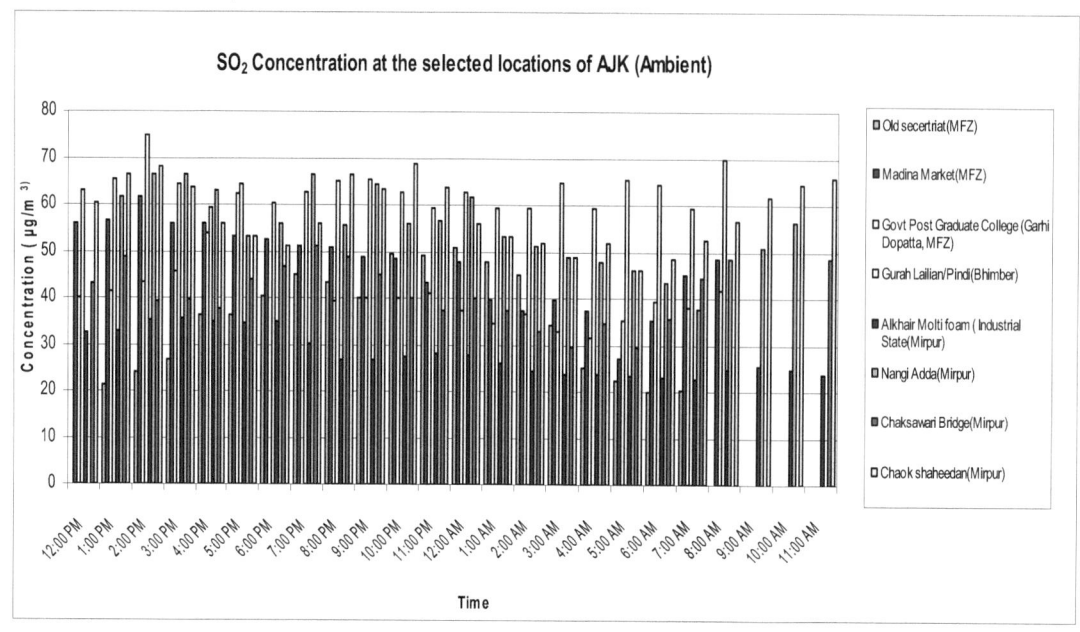

Fig 6.4

6.3 Carbon Monoxide (CO)

Carbon monoxide (CO) is a colorless, odorless and tasteless gas. It is formed by the incomplete combustion of fuels containing carbon. The main outdoor source of carbon monoxide is motor vehicles. Industrial sources include steel plants, foundries, and oil refining and chemical manufacturing facilities.

6.3.1 Health and Environmental Impacts of CO

CO can cause harmful health effects by reducing oxygen delivery to the body's organs (like the heart, brain) and tissues. The health threat from lower levels of CO is most serious for those who suffer from heart disease, like angina, clogged arteries, or congestive heart failure. High levels of CO can affect even healthy people. People who breathe high levels of CO can develop vision problems, reduced ability to work or learn reduced manual dexterity, and difficulty in performing complex tasks. At extremely high levels, CO is poisonous and can cause death. CO contributes to the formation of smog, ground-level ozone which can trigger serious respiratory problems.

6.3.2 Monitoring results of CO

CO Concentration (mg.m^{-3}) Muzaffarabad

S.No	Site Name	Min	Max	Average
1	Old Secretariat	2.4	8.8	5.2
2	Madina market	2.5	7.2	4.3
3	AIMS Hospital	9.9	12.7	11.1
4	Govt Girls High School (Sehli Sarkar)	7.8	8.6	8.2
5	Mujhoi (Household, Garhi Dopatta)	5.3	6.1	5.7
6	Govt Post Graduate College	3.7	4.7	4.1

CO Concentration (mg.m^{-3}) Bhimber

S.No	Site Name	Min	Max	Average
1	Gurah Lailian/Pindi	3.6	6.0	4.2
2	Govt. Dispensary	8.7	10.9	9.7

CO Concentration (mg.m^{-3}) Mirpur

S.No	Site Name	Min	Max	Average
1	Zahoor Food Industry (Old Industrial State)	9.4	11.7	10.0
2	Alkhair Molti foam (Industrial State)	2.8	4.9	3.6
3	Nangi Adda	2.9	5.2	3.8
4	Household	8.5	10.9	9.8
5	Chaksawari Bridge)	2.4	5.3	3.6
6	Chaok shaheedan	2.4	4.5	3.6

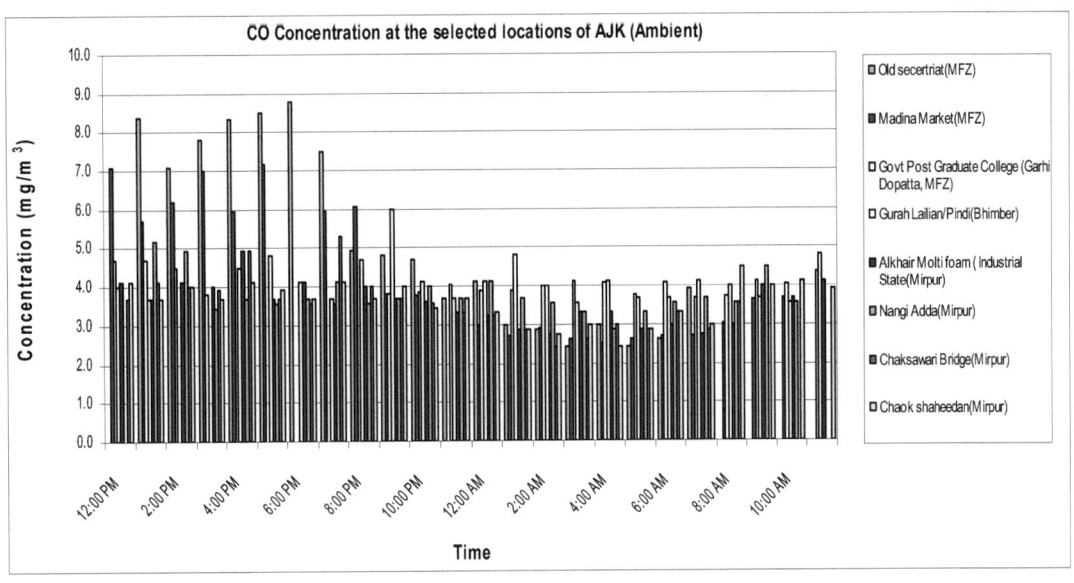

Fig 6.5

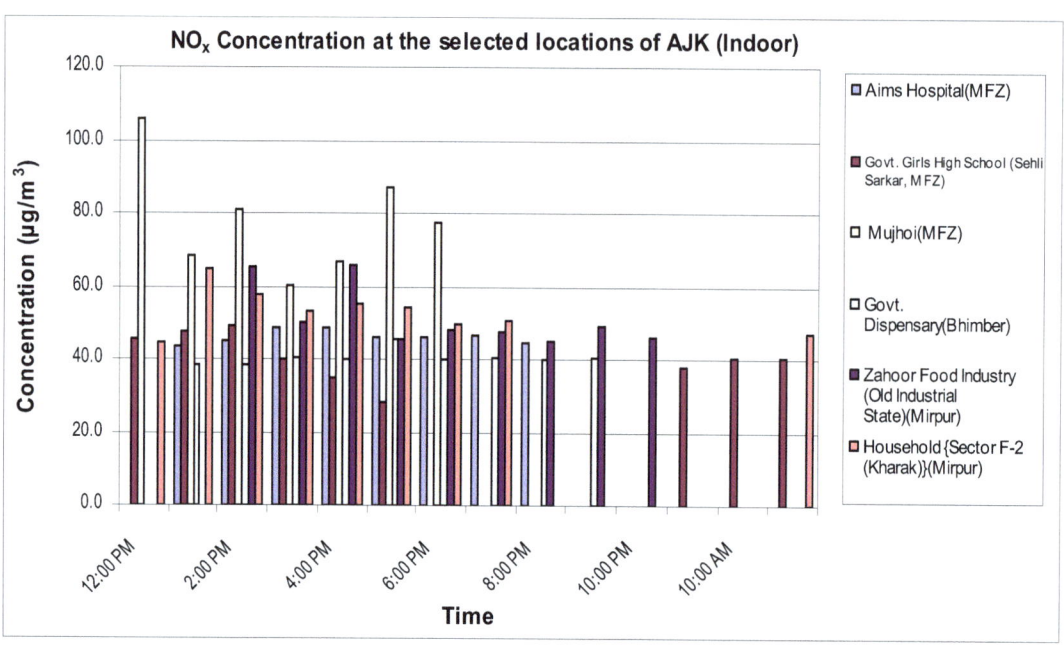

Fig 6.6

CO measured at the different sites of Muzaffarabad varied from 2.4 to 12.7 mg/m^3 and the average values recorded at different sites varied from 4.1 to 11.1 mg/m^3. All the values recorded at different location of Muzaffarabad area within the US EPA 10 mg/m^3 (8 hours) and 40 mg/m^3 10 mg/m^3 WHO and 10 mg/m^3 NEQS, respectively.

In Bhimber the hourly values varied from 3.6 to 10.9 mg.m^{-3} while the site averages are 4.2 and 9.7 mg.m^{-3} respectively. While in Mirpur it 2.8 to 11.7 mg.m^{-3} while the average values varied from 3.6 to 10.0 mg.m^{-3}. Overall CO level in AJK is lower than the 10 mg.m^{-3} of the NEQS values.

In a study in Karachi at 13 sites by Ghauri et al. (1992a) CO was in the range 10.4–11.5 mg/m^3. The first measurements of CO in Karachi were undertaken in 1969 as part of a survey at 26 road locations (Beg 1990).

Concentrations were in the range 6–23 mg/m^3 near the roadside and 12–41 mg/m^3 in the center of the road during traffic congestion. In 1983, a survey from January to June showed levels of 12–23 mg/m^3; but by 1988 CO concentrations had increased and 10-h means were in the range 2–57 mg/m^3 with short-term concentrations up to 107 mg/m^3 near heavy trafficked sites Beg (1990).

Siddique et al. (2005a) reported mean daily levels of CO for wood use and natural gas were 24 and 5 ppm while the levels of $PM_{2.5}$ were 12 and 0.25 mg/m^3, respectively. However, during cooking periods in the kitchens using biofuels, a sharp rise in concentration of CO (150 ppm) and $PM_{2.5}$ (300 mg/m^3) was seen. Colbeck et al. (2008) reported the results of an investigation on indoor air quality at rural and urban areas of Pakistan. Measurements were made of particulate mass (PM_{10}, $PM_{2.5}$, and PM_1), number concentration and bioaerosols in different micro-environments PM_{10} concentrations up to 8,555 µg/m^3 were observed inside kitchens where biofuels were burnt.

6.4 Carbon Dioxide (CO_2)

Carbon dioxide (CO_2) is a chemical compound composed of two oxygen atoms covalently bonded to a single carbon atom. It is a gas at standard temperature and pressure and exists in Earth's atmosphere in this state. CO_2 is a trace gas comprising 0.039% of the atmosphere. Carbon dioxide is used by plants during photosynthesis to make sugars, which may either be consumed in respiration or used as the raw material to produce other organic compounds needed for plant growth and development. It is produced during respiration by plants, and by all animals, fungi and microorganisms that depend either directly or indirectly on plants for food. It is thus a major component of the carbon cycle. Carbon dioxide is generated as a by-product of the combustion of fossil fuels or the burning of vegetable matter. Amounts of carbon dioxide are emitted from volcanoes and other geothermal processes. Carbon dioxide in the Earth's atmosphere is at a concentration of 391 ppm by volume. Atmospheric concentrations of carbon dioxide fluctuate slightly with the change of the seasons, driven primarily by seasonal plant growth in the Northern Hemisphere. Concentrations of carbon dioxide fall during the northern spring and summer as plants consume the gas, and rise during the northern autumn and winter as plants go dormant, die and decay. Carbon dioxide content in fresh air (averaged between sea-level and 10 hPa level, i.e. about 30 km altitude) varies between 0.036% (360 ppm) and 0.039%

(390 ppm), depending on the location. Prolonged exposure to moderate concentrations can cause acidosis and adverse effects on calcium phosphorus metabolism resulting in increased calcium deposits in soft tissues. Carbon dioxide is toxic to the heart and causes diminished contractile force. Toxicity and its effects increase with the concentration of CO_2, here given in volume percent of CO_2 in the air:

- 1% can cause drowsiness with prolonged exposure.
- At 2% it is mildly narcotic and causes increased blood pressure and pulse rate, and causes reduced hearing.
- At about 5% it causes stimulation of the respiratory centre, dizziness, confusion and difficulty in breathing accompanied by headache and shortness of breath. Panic attacks may also occur at this concentration. At about 8% it causes headache, sweating, dim vision, tremor and loss of consciousness after exposure for between five and ten minutes. Due to the health risks associated with carbon dioxide exposure, the U.S. Occupational Safety and Health Administration says that average exposure for healthy adults during an eight-hour work day should not exceed 5,000 ppm (0.5%).

CO_2 Concentration (mg.m^{-3}) in Muzaffarabad

S.No	Site Name	Min	Max	Average
1	Old Secretariat	578	749	641
2	Madina Market	565	681	586
3	AIMS Hospital	591	668	622
4	Govt Girls High School (Sehli Sarkar)	614	716	666
5	Mujhoi (Household, Garhi Dopatta)	636	746	690
6	Govt Post Graduate College	578	653	602

CO_2 Concentration (mg.m^{-3}) in Bhimber

S.No	Site Name	Min	Max	Average
1	Gurah Lailian/Pindi	585	716	660
2	Govt. Dispensary	678	877	737

CO_2 Concentration (mg.m^{-3}) in Mirpur

S.No	Site Name	Min	Max	Average
1	Zahoor Food Industry (Old Industrial State)	665	697	679
2	Alkhair Molti foam(Industrial State)	593	812	691
3	Nangi Adda	642	957	806
4	Household	668	740	687
5	Chaksawari Bridge)	585	713	627
6	Chaok shaheedan	630	964	777

CO_2 measured at the different sites of Muzaffarabad varied from 565 to 749 mg.m^{-3} and the average values recorded at different sites varied from 586 to 690 mg.m^{-3}. In Bhimber the hourly CO_2 values varied from 585 to 877 mg.m^{-3} while the site averages are 660 and 737 mg.m^{-3} respectively. While in Mirpur CO_2 values were 585 to 964 mg.m^{-3} while the average values varied from 627 to 806 mg.m^{-3} respectively.

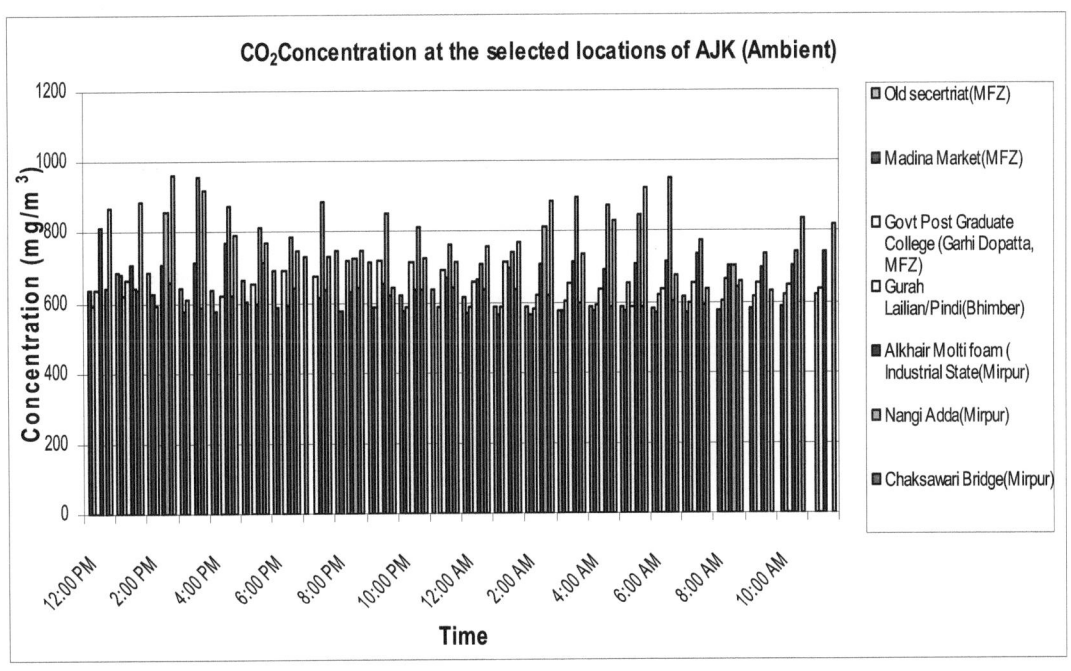

Fig 6.7

6.5 Particulate Matter ($PM_{10}/PM_{2.5}$)

Material suspended in the air in the form of minute solid particles or liquid droplets, especially when considered as an atmospheric pollutant.

6.5.1 Sources

Unlike the individual gaseous pollutants which are single, well-defined substances, particles (PM_{10}) in the atmosphere are composed of a wide range of materials arising from a variety of sources. Concentrations of PM_{10} comprise, primary particles arising from combustion sources; and secondary particles, mainly sulphate, nitrate formed by chemical reactions in the atmosphere, coarse particles, suspended soils, dusts, sea-salt, biological particles and particles from the construction work. The relative contribution of each source type varies from day to day depending on meteorological conditions, quantities of emissions from mobile and static sources. $PM_{2.5}$ material is primarily formed from chemical reactions in the atmosphere and through fuel combustion (e.g., motor vehicles, power generation, industrial facilities, residential fire places, wood stoves and agricultural burning).

6.5.2 Health effects

Particulate air pollution is associated with a range of impacts on health including effects on the respiratory and cardiovascular systems including asthma.

6.6 Monitoring results of PM10 and PM2.5

PM_{10} and $PM_{2.5}$ Concentration ($\mu g.m^{-3}$) in Muzaffarabad

S.No	Site Name	PM 10	PM 2.5
1	Old Secretariat	233	128
2	Madina market	111	77
3	AIMS Hospital	173	91
4	Govt Girls High School (Sehli Sarkar)	27	16
5	Mujhoi (Household, Garhi Dopatta)	130	64
6	Govt Post Graduate College	139	64

PM$_{10}$ and PM$_{2.5}$ Concentration (µg.m^{-3}) in Bhimber

S.No	Site Name	PM 10	PM 2.5
1	Gurah Lailian/Pindi	195	61
2	Govt. Dispensary	284	116

PM$_{10}$ and PM$_{2.5}$ Concentration (µg.m^{-3}) in Mirpur

S.No	Site Name	PM 10	PM 2.5
1	Zahoor Food Industry (Old Industrial State)	158	36
2	Alkhair Molti foam(Industrial State)	88	25
3	Nangi Adda	154	120
4	Household	147	111

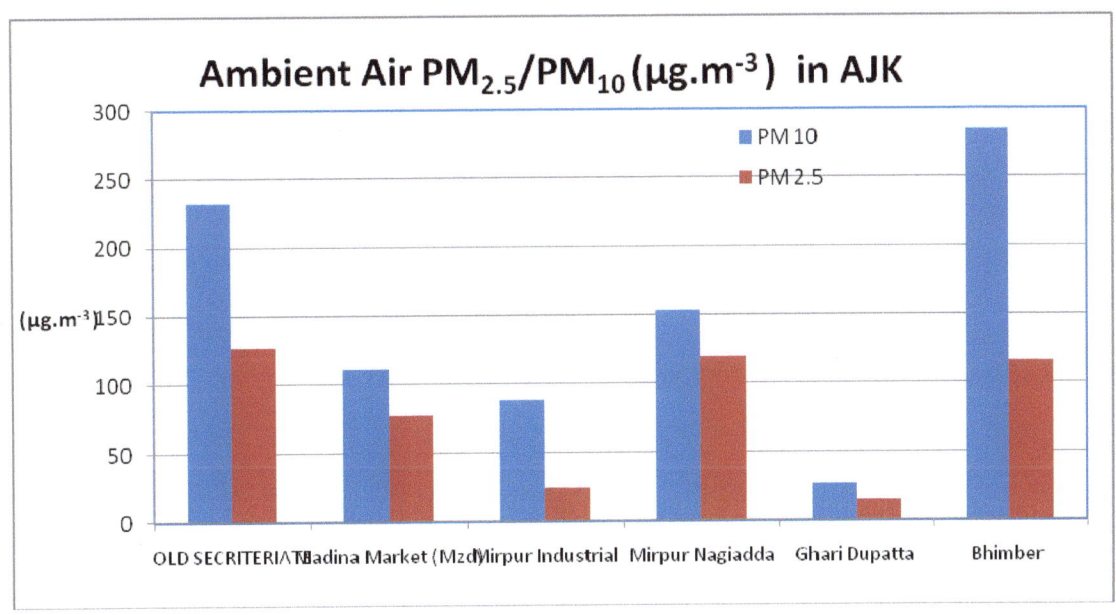

Fig 6.8

Site average PM$_{10}$ measured at different locations in Muzaffarabad varied from 27 to 233 µg.m^{-3} and PM$_{2.5}$ varied from 16 to 128µg.m^{-3} respectively. Similarly PM$_{10}$ in Bhimber values were 195 and 284 µg.m^{-3} and PM2.5 were 61 to 116 µg.m^{-3} respectively. In Mirpur PM$_{10}$ varied from 88 to 158 and PM$_{2.5}$ values varied from 25 to 120 µg.m^{-3} respectively.

All the values recorded at different locations of AJK are within the US EPA, NEQS value of 250 µg/m^3.

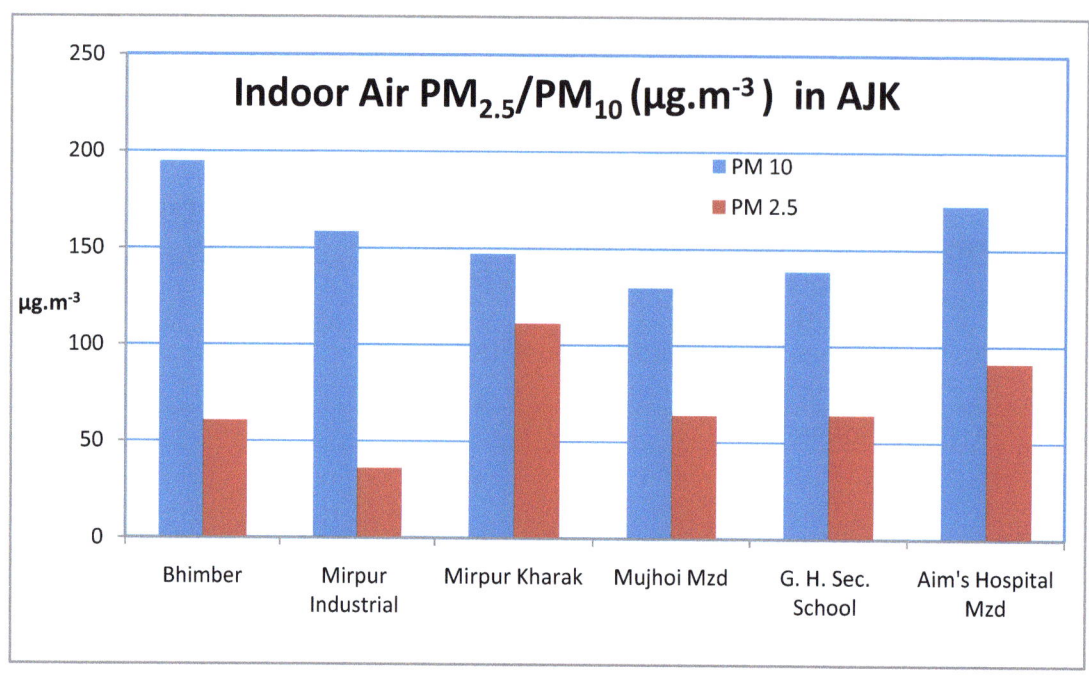

Fig 6.9

If the results of the current study is compared with some other studies conducted in Pakistan, for example Stone et al 2009 reported particulate matter ($PM_{2.5}$ and PM_{10}) collected in Lahore, Pakistan from 12 January 2007 to 19 January 2008 on the campus of the University of Engineering and Technology on the roof of the Institute for Environmental Engineering and Research at a height of 10m. According to the results annual average concentration of $PM_{2.5}$ of 194±94 µg m^{-3} and PM_{10} was 336±135 µg.m^{-3}. Coarse aerosol ($PM_{10-2.5}$) was dominated by crustal sources like dust (74±16%, annual average ±one standard deviation) whereas fine particles were dominated by carbonaceous aerosol (organic matter and elemental carbon, 61±17%) (Stone, Schauer et al. 2010). These results are higher from the values measured in the present study. Shah et al. 2004 reported particulate matter (PM) collected from two sampling stations in Islamabad, Pakistan; using high volume air sampler on glass fiber filters for 8–12 h on daily basis. Particle size fractions were categorized as, 2.5, 2.5–10, 10–100 and .100 mm. The results from two stations indicated average airborne lead concentrations of 0.505 and 0.185 µg.m^{-3}. $PM_{2.5}$ and PM_{10} were found to constitute the local atmosphere in comparable proportions (Shah, Shaheen et al. 2004).

The 2007 annual average PM_{10} mass concentration was 340 µg m^{-3}, which is well above the WHO guideline of 20 µg m^{-3} (von Schneidemesser, Stone et al. 2010). Zhang et al., (2008) reported high PM_{10} mass concentrations, averaging of 459 µg/m^3.

According to Lodhi et al., (2009) $PM_{2.5}$ concentration during November 2005 to January 2006 ranged from 53 µg m^{-3} to 476 µg m^{-3} with the mean value of 191 ± 90 µg m^{-3}. During February 2006 to March 2006, $PM_{2.5}$ level ranged from 32 to 400 µg m^{-3} with the mean value of 143 ±100 µg m^{-3}.

Zhang et al., (2008) reported high PM_{10} mass concentrations, averaging 459 µg/m^3.

Shah & Shaheen (2008) report finding of their research to assess concentration levels and sources of selected metals in TSP and their seasonal variability; and the role of climatic conditions on TSP load and metal concentrations. According to results TSP varied from a minimum of 41.8 to a maximum of 977 µg/m^3, with a mean value of 164 µg/m^3 (Shah and Shaheen 2008).

Raja, Biswas et al. (2010) calculated samples of airborne particulate matter ($PM_{2.5}$) collected at a site in Lahore, Pakistan from November 2005 to January 2006. The highest $PM_{2.5}$ value was 476µg m^{-3} measured on December 21, 2005 and the lowest concentration was 53µg m^{-3} measured on November 30, 2005. They concluded that most of the high $PM_{2.5}$ mass values measured in Lahore were likely due to emissions from diesel trucks, two-stroke vehicles, and possibly from vehicle-emitted primary sulfate that has been included in the "secondary aerosol" factor.

Qadir and Zaidi (2006) have reported total suspended particulate matter using standard gravimetric technique. The trace elemental composition in the atmosphere of Faisalabad was studied by using instrumental neutron activation analysis (INAA) and atomic absorption spectrometry (AAS). The atmospheric mass concentration of total suspended particulates (TSP) on working days at Faisalabad ranged from 467 to 600 µg.m^{-3} with an average of 550 µg.m^{-3} (Qadir and Zaidi 2006).

Qadir discusses the existing air quality in urban centers of Pakistan, current activities for controlling this menace, their impact on urban air quality and future suggestion to include transport management for reduction of urban air pollution in the context of a developing country like Pakistan. The current air pollution scenario in Pakistan' cities are transport and energy sector contributes nearly one half of the NOx, two-thirds of CO, and about one half of hydrocarbon emissions in the industrialized countries. The major constituents of urban air pollution consist of Total SPM, PM_{10}, $PM_{2.5}$, Lead, SO_2, CO, O_3, NO, NO_2, non-methane hydrocarbons, and the levels are higher than normal. Recent surveys carried out in the country using mobile units (for example one carried out with the assistance of JICA and Pak EPA) revealed presence of very high levels of suspended particulate matter (SPM12) in major cities (about 6 times higher than the World Health Organization's guidelines). In Lahore, Rawalpindi and Karachi levels of Carbon monoxide, oxides of Nitrogen and Sulphur dioxide were also found in high concentration in other studies (Qadir 2002).

Ali and Athar (2010) present the finding of ambient air quality monitoring carried out in Lahore City, Pakistan. The ambient air quality was monitored for criteria pollutants carbon monoxide (CO), nitrogen dioxide (NO_2), sulfur dioxide (SO_2), ozone (O3), particulate matter (TSP and PM_{10}), lead (Pb), and noise level at ten different locations of the city.

The concentration of nitrogen dioxide was found in the range of 8.2 to 24.5 ppb. The concentration of sulfur dioxide was found in the range of 6.7 to 27.3 ppb. The concentration of ozone was found in the range of 14.7–21.2 ppb. The concentration of total suspended particulate (TSP) was found very high in the range of 324–1334 µg/m^3 at different locations. The concentration of lead at different locations of the city was found in the range of 1.9 to 7.2 µg/m^3. The 24-h noise level average was observed in the range of 67.5 to 88.2 dB at different locations of the city (Ali and Athar 2010).

Ghauri et al, (1994) reports the results of an atmospheric pollution survey carried out at 13 sites in, Karachi, Pakistan, simultaneously from 0600h to 2100.The conclusion is that carbon monoxide levels exceeded in the ambient air were found to reach 9-10 ppm along the busy urban streets whereas CO_2 level exceeded 370 ppm in these areas. This survey indicates that NO_2 levels were exceeding U.S. ambient air quality standards. The surface ozone maximum around noon reached the levels of 40ppb and 50 ppb respectively compared to upwind coastal Sites level of 25 ppb. The authors concludes contaminants that exist throughout urban communities or in localized areas (such as M.A.Jinnah Road, Sadder, Korangi) of the Karachi metropolitan area is of great importance to health (Ghauri, Salam et al. 1994).

Ghauri et al (2007) reports the results of a comprehensive study conducted by SUPARCO during 2003-2004 for a yearlong baseline air quality study in country's major urban areas (Karachi, Lahore, Quetta, Rawalpindi, Islamabad and Peshawar). The objective of this study was to establish baseline levels and behavior of airborne pollutants in urban centers with temporal and spatial parameters. This study reveals that the highest concentrations of CO were observed at Quetta (14ppm) while other pollutants like SO_2 (52.5 ppb), NO_x (60.75 ppb) and O_3 (50 ppb) were higher at Lahore compared to other urban centers like Karachi and Peshawar, etc. The maximum particulate (TSP) and PM_{10} levels were observed at Lahore (996 µg/m^3 and 368 µg/m^3 respectively), Quetta (778 µg/m^3, 298 µg/m^3) and in Karachi (410 µg/m^3, 302 µg/m^3).these hourly and daily mean concentrations were compared with USEPA and NEQS prescribed limit (Ghauri, Lodhi et al. 2007).High levels of particulate matter (PM2.5, PM10) were reported in six Asian cities (Bandung, Bangkok, Beijing, Chennai, Manila, and Hanoi) by Oanh et al. (2006) within the framework of the Asian regional air pollution research network. The average concentrations of $PM_{2.5}$ and PM_{10} were in the range 44–168 and 54–262µg/m^3 in the dry season, and 18–104 and 33–180µg/m^3 in the wet season, respectively (Oanh and Zhang 2004). Many Government departments and international organizations have identified degradation of ambient air quality as a major environmental concern in Pakistan. Industrial pollution, suspended particulates, indoor air pollution, and increasing traffic trends were reported as key sources affecting ambient air quality in the country (Pak-EPA 2005), Pakistan Economic Survey Report 2006–2007; World Bank (2006b); Pakistan Millennium Development Goals Report 2005). In Lahore during 1978–1980 under the Global Environment Monitoring System (GEMS). The annual mean level of suspended particulate matter (SPM) at the commercial city center was 332µg/m^3 during 1978. At the suburban residential site, a concentration of 749 and 690µg/m^3 was reported during the period of 1979 and 1980, respectively (World Health Organization 1984).The average levels of PM_{10} measured (Hashmi and Khan 2003) with a mobile monitoring laboratory at the Sindh Industrial Trading Estate and Korangi Industrial Area (Karachi) were 176.5 and 147.2µg/m^3, respectively. The hourly average PM10 concentration at Port Qasim in Karachi for 7 days during November was 123.49µg/m^3 (Hashmi, Shaikh et al. 2005).

An investigation (Rajput, Ahmad et al. 2005) on TSP levels and its chemical composition in industrial and residential areas of Islamabad during 1995 depicted that the levels of TSP in the industrial area (297µg/m^3) were more than double those of the residential area (133µg/m^3).A study conducted by the Environment Agency of Pakistan in Quetta reported levels of particulate matter at four different locations (two kerbside, one industrial, one residential). The concentrations of TSP, PM_{10}, and $PM_{2.5}$ varied between 385–1,778µg/m^3, 126–709µg/m^3 and 104–222µg/m^3, respectively (EPA), published in year 2007. Colbeck, Nasir et al. (2010) reported the results of a study carried out on indoor/outdoor particulate pollution in rural and urban residential environments.

In the kitchens of rural areas using biomass fuel, the 24-h average indoor concentration of PM10, PM2.5, and PM1 was 1,581±2,003, 1,169±1,489, and 913±992 µg/m^3, respectively. In rural living rooms, for the same time period and particle size, the concentrations were 953±641, 603±421, and 548±400 µg/m^3, respectively. On the other hand, in the urban living room, the 24-h average indoor mass concentrations for the same size fractions were 533±641, 402±641, and 362±641 µg/m^3, respectively. From the review of all these studies it is concluded that PM_{10}/$PM_{2.5}$ is on lower side as compared to other parts of Pakistan.

6.7 Noise and its Health effects

Sound or a sound that is loud, unpleasant, unexpected, or undesired. Elevated workplace or other noise levels can cause hearing impairment, hypertension, ischemic heart disease, annoyance, sleep disturbance. Noise exposure has also been known to induce hypertension, and other cardiovascular impacts. Beyond these effects elevated noise levels can create stress and increased workplace accident rates.

Noise Level (dB) in Muzaffarabad

S.No	Site Name	Min	Max	Average
1	Old Secretariat	41.4	77.3	61.9
2	Madina market	62.4	72.0	68.9
3	AIMS Hospital	45.0	55.0	51.1
4	Govt Girls High School (Sehli Sarkar)	43.7	47.5	45.7
5	Mujhoi (Household, Garhi Dopatta)	44.5	53.8	48.5
6	Govt Post Graduate College	41.4	77.3	61.9

Noise Level (dB) in Bhimber

S.No	Site Name	Min	Max	Average
1	Gurah Lailian/Pindi	56.3	128.2	95.5
2	Govt. Dispensary	64.6	90.7	75.8

Noise Level (dB) in Mirpur

S.No	Site Name	Min	Max	Average
1	Zahoor Food Industry (Old Industrial State)	48.0	57.3	52.7
2	Alkhair Molti foam(Industrial State)	47.0	71.3	55.7
3	Nangi Adda	53.5	76.8	67.9
4	Household	42.0	46.8	45.4
5	Chaksawari Bridge)	56.3	67.0	61.3
6	Chaok shaheedan	60.8	74.5	69.0

Noise Levels in Muzaffarabad varied from 41.4 to 77.3 dB while the site average values varied from 45.7 to 68.9 dB respectively. In Bhimber it varied from 56.3 to 128.2 while the site average was 75.8 and 95.5 dB respectively. Noise levels in Mirpur varied from 42.0 to 76.8 while site averages varied from 45.4 to 69.0 dB respectively. Noise levels were with in the NEQS day time values of (65 Residential, Commercial Area 70 and

80 Industrial Area values respectively. Overall average Indoor noise levels were 51.8 and out door noise levels were 60 well with in the daytime limits examined.

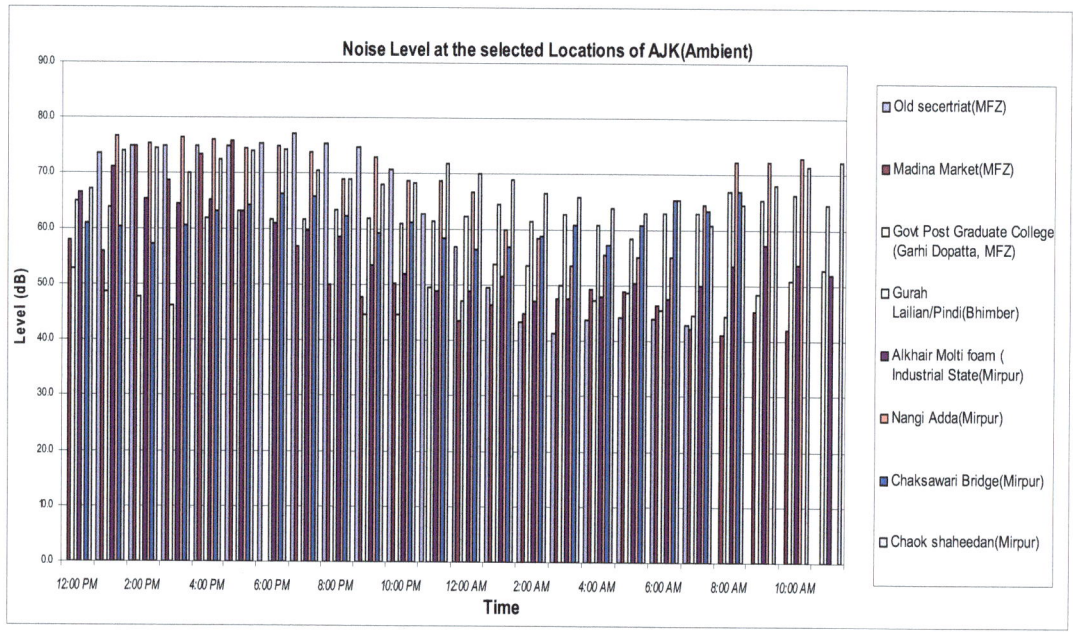

Fig 6.10

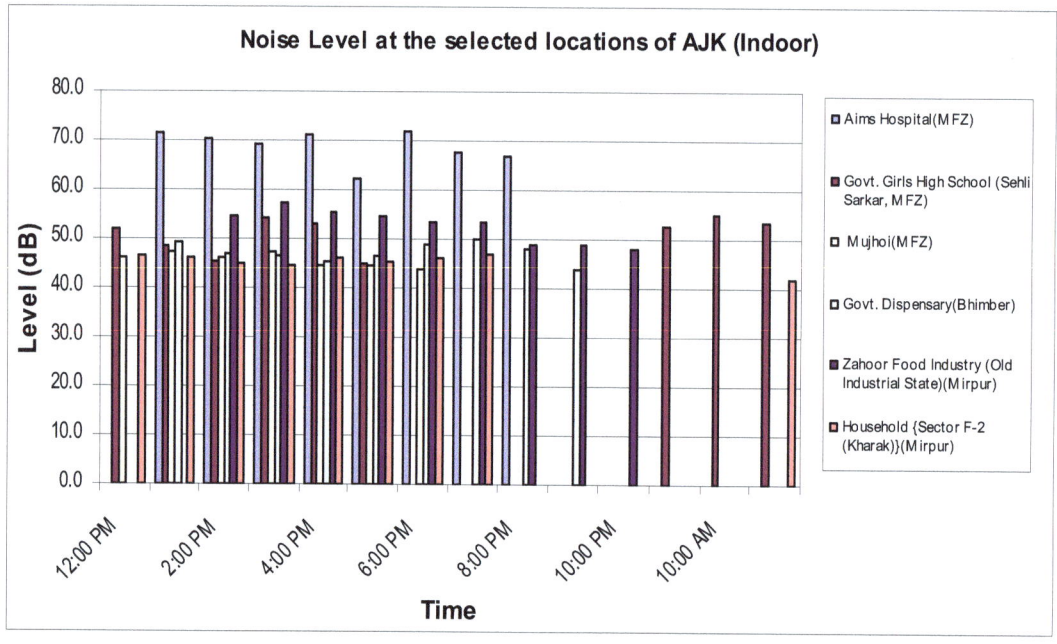

Fig 6.11

If the results of the current study are compared with some other studies conducted in Pakistan for example Wilson and Nicol (1994) have climatic variation in noise tolerance and the interaction with the thermal environment. Two respondents (experiencing sound of 74 dB and 71 dB) judged the noise acceptable while a third (experiencing 69 dB) found it rather noisy. The experiment was performed in Karachi (25m from a main road in Karachi, was 6 meters wide) and naturally ventilated in winter though air-conditioned in summer.

The main source of noise in the office was the street noise as the noise level dropped from 65-68 dB on day 1 with 2 windows open to 62-64 dB on day 2 with 1 window open and to 60-61 dB with the windows closed. Indoor noise was measured 2m from the window. Outdoor noise was measured for about an hour on day 2, 1 m from the windows and a level of about 72-74 dB recorded (Wilson and Nicol 1994). Thus the levels recorded in the different locations of AJK are lower than the above mentioned study.

6.8 Ozone

Ozone (O_3), or Trioxygen, is a triatomic molecule, consisting of three oxygen atoms. It is an allotrope of oxygen that is much less stable than the diatomic allotrope (O_2). Ozone in the lower atmosphere is an air pollutant with harmful effects on the respiratory systems of animals and will burn sensitive plants; however, the ozone layer in the upper atmosphere is beneficial, preventing potentially damaging ultraviolet light from reaching the Earth's surface. Ozone is present in low concentrations throughout the Earth's atmosphere. Low level ozone (or tropospheric ozone) is regarded as a pollutant by the WHO (World Health Organization 2010) and the United States Environmental Protection Agency (EPA). It is not emitted directly by car engines or by industrial operations, but formed by the reaction of sunlight on air containing hydrocarbons and nitrogen oxides that react to form ozone directly at the source of the pollution or many kilometers down wind.

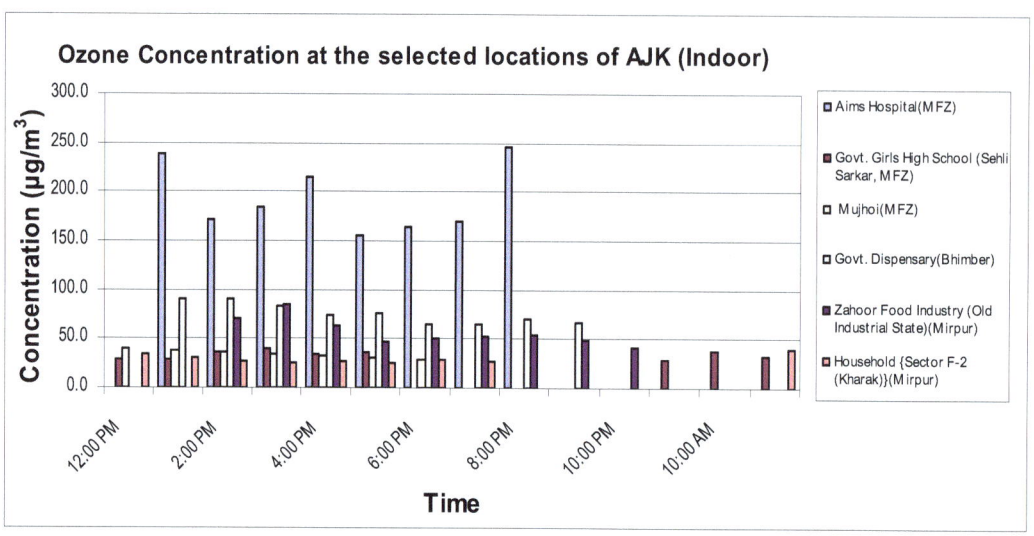

Fig 6.12

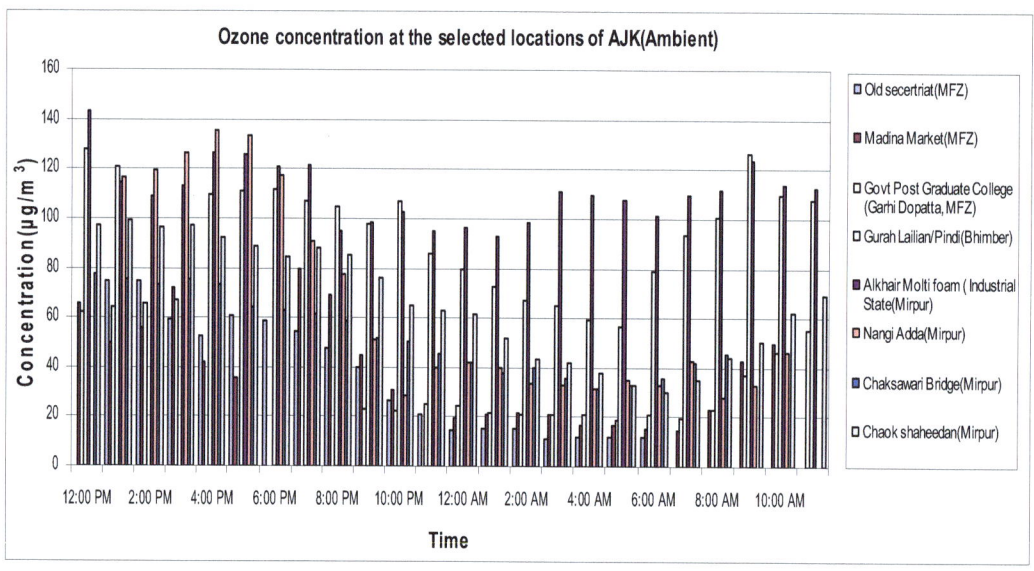

Fig 6.13

Ozone reacts directly with some hydrocarbons such as aldehydes and thus begins their removal from the air, but the products are themselves key components of smog. Ozone photolysis by UV light leads to production of hydroxyl radical OH and this plays a part in the removal of hydrocarbons from the air, but is also the first step in the creation of components of smog such as peroxyacyl nitrates which can be powerful eye irritants.

The atmospheric lifetime of tropospheric ozone is about 22 days; its main removal mechanisms are being deposited to the ground, the above mentioned reaction giving OH, and by reactions with OH and the peroxy radical HO_2. (Stevenson, Dentener et al. 2006).

Ozone Levels ($\mu g.m^{-3}$) in Muzaffarabad

S.No	Site Name	Min	Max	Average
1	Old Secretariat	11.4	75.0	35.8
2	Madina market	14.7	79.5	38.5
3	AIMS Hospital	155.4	246.6	193.2
4	Govt Girls High School (Sehli Sarkar	28.5	39.8	33.8
5	Mujhoi (Household, Garhi Dopatta)	29.6	39.2	34.2
6	Govt Post Graduate College	18.7	67.4	34.8

Ozone Levels ($\mu g.m^{-3}$) in Bhimber

S.No	Site Name	Min	Max	Average
1	Gurah Lailian/Pindi	56.3	128.2	95.5
2	Govt. Dispensary	64.6	90.7	75.8

Ozone Levels ($\mu g.m^{-3}$) in Mirpur

S.No	Site Name	Min	Max	Average
1	Zahoor Food Industry (Old Industrial State)	42.4	85.8	57.2
2	Alkhair Molti foam(Industrial State)	92.7	143.4	110.7
3	Nangi Adda	27.8	135.9	65.2
4	Household	25.7	39.4	29.9
5	Chaksawari Bridge)	31.2	77.5	52.9
6	Chaok shaheedan	30.4	98.9	66.3

Ozone levels in Muzaffarabad varied from 11.4 to 246.6 ($\mu g.m^{-3}$) while site averages varied from 34.2 to 193.2 $\mu g.m^{-3}$ respectively. Ozone levels at Bhimber varied from 56.3 to 128.2 $\mu g.m^{-3}$ respectively, while the site averages were 75.8 and 95.5 $\mu g.m^{-3}$ respectively. In Mirpur ozone levels varied from 25.7 to 143.4 $\mu g.m^{-3}$ while the site averages were 29.9 to 110.7 $\mu g.m^{-3}$ respectively. The results of this are with in the NEQS proposed value of 180 $\mu g/m^3$ (1Hour) value. According to the GEMS study, the daily average of O_3 in Karachi was in the range 36–50 $\mu g/m^3$ (WHO/UNEP 1992). Ghauri, Salam et al. (1991) reported measurement of O_3 during 1986–1988 at three sites (one upwind and two downwind) in Karachi.

Yousufzai, Hashmi et al. (2000) performed continuous measurements of O3 and found the lowest level of ozone was 15μg/m^3 rising to 38μg/m^3 during the day. According to Ghauri et al. (2007), the 48-h mean concentration of ozone was highest in Karachi (50 μg/m^3), followed by Quetta (48 μg/m^3), Peshawar (46 μg/m^3), Lahore (44 μg/m^3), Islamabad (36 μg/m^3), and Rawalpindi (34 μg/m^3). Generally, maximum levels were found in the afternoon and peak concentrations were recorded during the summer.

6.9. Volatile Organic Compounds

A volatile organic compound (VOCs) refers to organic chemical compounds which have significant vapor pressures and which can affect the environment and human health. VOCs are numerous, varied, and ubiquitous. Although VOCs include both man-made and naturally occurring chemical compounds, it is the anthropogenic VOCs that are regulated, especially for indoors where concentrations can be highest. VOCs are typically not acutely toxic but have chronic effects. Because the concentrations are usually low and the symptoms slow to develop, analysis of VOCs and their effects is a demanding area.

Major sources of man-made VOCs are solvents, especially paints and protective coatings. Solvents are required to spread a protective or decorative film. Approximately 12 billion liters of paints are produced annually. Industrial use of fossil fuels produces VOCs either directly as products (e.g. gasoline) or indirectly as byproducts (e.g. automobile exhaust). Many building materials such as paints, adhesives, wall boards, and ceiling tiles slowly emit formaldehyde, which irritates the mucous membranes and can make a person irritated and uncomfortable. Formaldehyde emissions from wood are in the range of 0.02 – 0.04 ppm. Relative humidity within an indoor environment can also affect the emissions of formaldehyde. High relative humidity and high temperatures allow more vaporization of formaldehyde from wood-materials. There are also many sources of VOCs in office buildings, which include new furnishings, wall coverings, and office equipment such as photocopy machines which can off-gas VOCs into the air. MTBE was banned in the US around 2004 in order to limit further contamination of drinking water aquifers.

6.9.1 Health affects

Respiratory, allergic, or immune effects in infants or children are associated with indoor VOCs and other indoor air pollutants. Some VOCs, such as styrene and limonene, can react with nitrogen oxides or with ozone to produce new oxidation products and secondary aerosols, which can cause sensory irritation symptoms. Unspecified VOCs are important in the creation of smog.

6.10. Monitoring results of BTEX

BTEX concentrations measured at different locations of Muzaffarabad and accordingly benzene varied from 0.3 to 14.0 (indoor) and the site average values varied from 0.9 to 12.2ppb respectively. In Bhimber Benzene varied from 2.1 to 9.6 ppb and the site average values were 3.4 to 7.7 ppb respectively. In Mirpur it varied from 1.2 to 10.3 and the site averages were 1.3 to 9.5 ppb, respectively.

BTEX Concentration (ppb) in Bhimber

Site Name	Pollutants	Min	Max	Average
Gurah Lailian/Pindi	Benzene(ppb)	4.1	9.6	7.7
	Toluene (ppb)	1.2	1.6	1.3
	Ethylbenzene(ppb)	0.0	0.8	0.2
	Xylene (ppb)	0.0	0.2	0.1
	M+P Xylene (ppb)	0.0	0.3	0.0
Govt. Dispensary	Benzene(ppb)	2.1	5.2	3.4
	Toluene (ppb)	1.2	1.6	1.3
	Ethylbenzene(ppb)	0.2	0.9	0.4
	Xylene (ppb)	0.1	0.5	0.3
	M+P Xylene (ppb)	0.0	0.5	0.2

BTEX Concentration (ppb) in Muzaffarabad

Site Name	Pollutants	Min	Max	Average
Old Secretariat	Benzene(ppb)	1.6	8.4	4.9
	Toluene (ppb)	1.2	3.4	2.1
	Ethylbenzene(ppb)	1.1	1.9	1.4
	Xylene (ppb)	1.3	4.2	2.2
	M+P Xylene (ppb)	1.2	3.4	1.7
Madina market	Benzene(ppb)	2.3	8.6	6.0
	Toluene (ppb)	1.0	3.6	2.3
	Ethylbenzene (ppb)	1.0	2.5	1.7
	Xylene (ppb)	0.2	1.6	0.8
	M+P Xylene (ppb)	0.1	1.9	0.7
AIMS Hospital	Benzene(ppb)	0.3	2.0	0.9
	Toluene (ppb)	0.0	0.3	0.1
	Ethylbenzene (ppb)	0.2	0.9	0.4
	Xylene (ppb)	0.2	0.6	0.3
	M+P Xylene (ppb)	0.1	0.6	0.3
Govt Girls High School (Sehli Sarkar	Benzene(ppb)	5.2	8.5	7.0
	Toluene (ppb)	2.1	3.5	2.6
	Ethylbenzene (ppb)	2.0	2.6	2.4
	Xylene (ppb)	2.3	3.6	3.2
	M+P Xylene (ppb)	1.0	1.9	1.5
Mujhoi (Household, Garhi Dopatta)	Benzene(ppb)	11.3	14.0	12.2
	Toluene (ppb)	3.0	7.8	6.1
	Ethylbenzene (ppb)	1.0	3.2	2.3
	Xylene (ppb)	2.5	4.2	3.3
	M+P Xylene (ppb)	2.1	2.9	2.5
Govt Post Graduate College	Benzene(ppb)	1.0	7.1	3.2
	Toluene (ppb)	1.0	4.2	2.5
	Ethylbenzene (ppb)	0.1	1.9	0.9
	Xylene (ppb)	0.1	2.3	0.5
	M+P Xylene (ppb)	0.1	1.9	0.9

BTEX Concentration (ppb) in Mirpur

Site Name	Pollutants	Min	Max	Average
Five Star Foam (New Industrial State)	Benzene(ppb)	1.2	1.6	1.3
	Toluene (ppb)	0.2	0.9	0.6
	Ethylbenzene(ppb)	0.2	0.6	0.4
	Xylene (ppb)	0.0	0.5	0.1
	M+P Xylene (ppb)	0.0	0.1	0.1
Nangi Adda	Benzene(ppb)	3.2	6.9	4.6
	Toluene (ppb)	0.1	1.6	0.8
	Ethylbenzene (ppb)	0.4	3.2	1.6
	Xylene (ppb)	0.0	0.9	0.3
	M+P Xylene (ppb)	0.2	1.7	0.8
Chaok Shaheedan	Benzene(ppb)	8.7	10.3	9.5
	Toluene (ppb)	4.0	8.6	6.2
	Ethylbenzene (ppb)	2.0	4.6	3.2
	Xylene (ppb)	1.5	2.6	2.2
	M+P Xylene (ppb)	1.0	1.6	1.3
Chaksawari Bridge	Benzene(ppb)	1.3	3.6	2.8
	Toluene (ppb)	0.0	0.3	0.2
	Ethylbenzene (ppb)	0.0	0.0	0.0
	Xylene (ppb)	0.0	0.0	0.0
	M+P Xylene (ppb)	0.0	0.0	0.0
Zahoor Food Industry (Old Industrial State)	Benzene(ppb)	1.8	5.3	3.0
	Toluene (ppb)	0.5	1.6	1.0
	Ethylbenzene (ppb)	0.0	0.6	0.2
	Xylene (ppb)	0.0	0.6	0.1
	M+P Xylene (ppb)	0.0	0.6	0.2
Household	Benzene(ppb)	3.1	9.5	6.1
	Toluene (ppb)	1.0	3.5	2.1
	Ethylbenzene (ppb)	0.0	2.6	0.4
	Xylene (ppb)	0.0	1.0	0.2
	M+P Xylene (ppb)	0.1	3.0	0.4

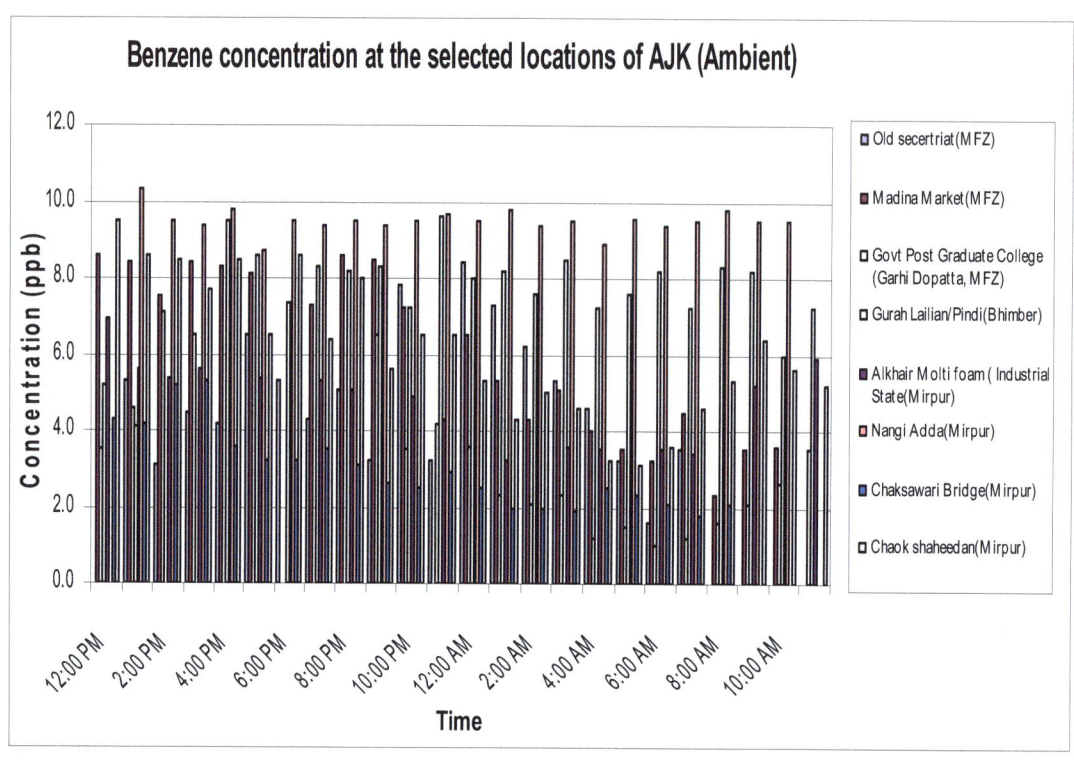

Fig 6.14

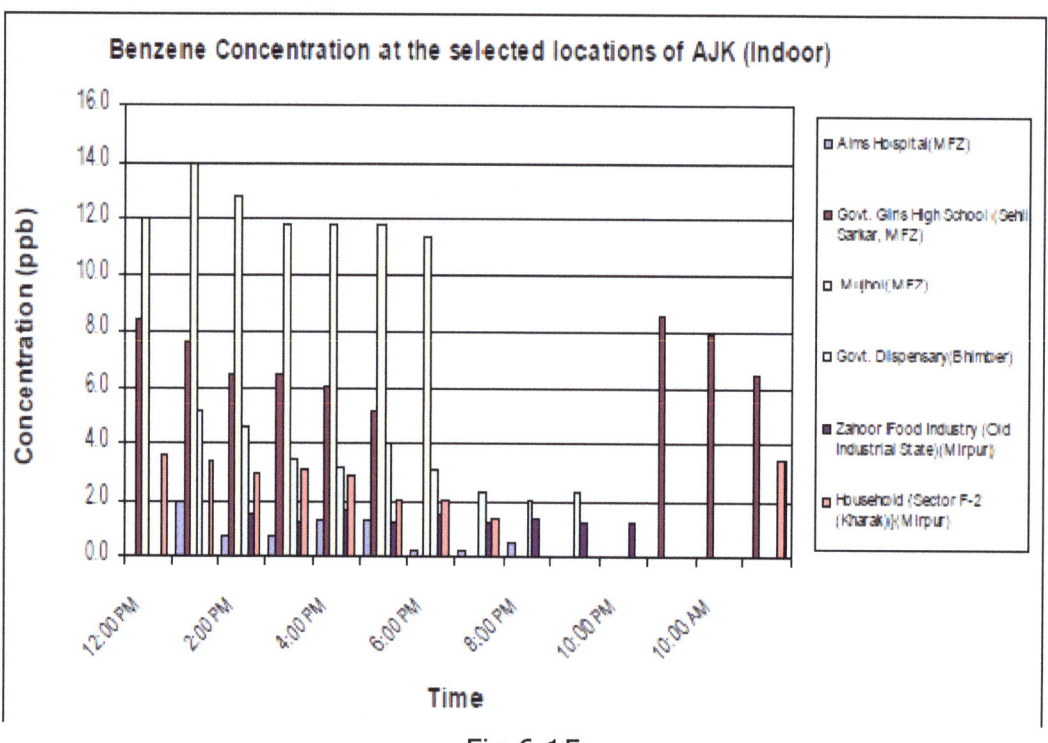

Fig 6.15

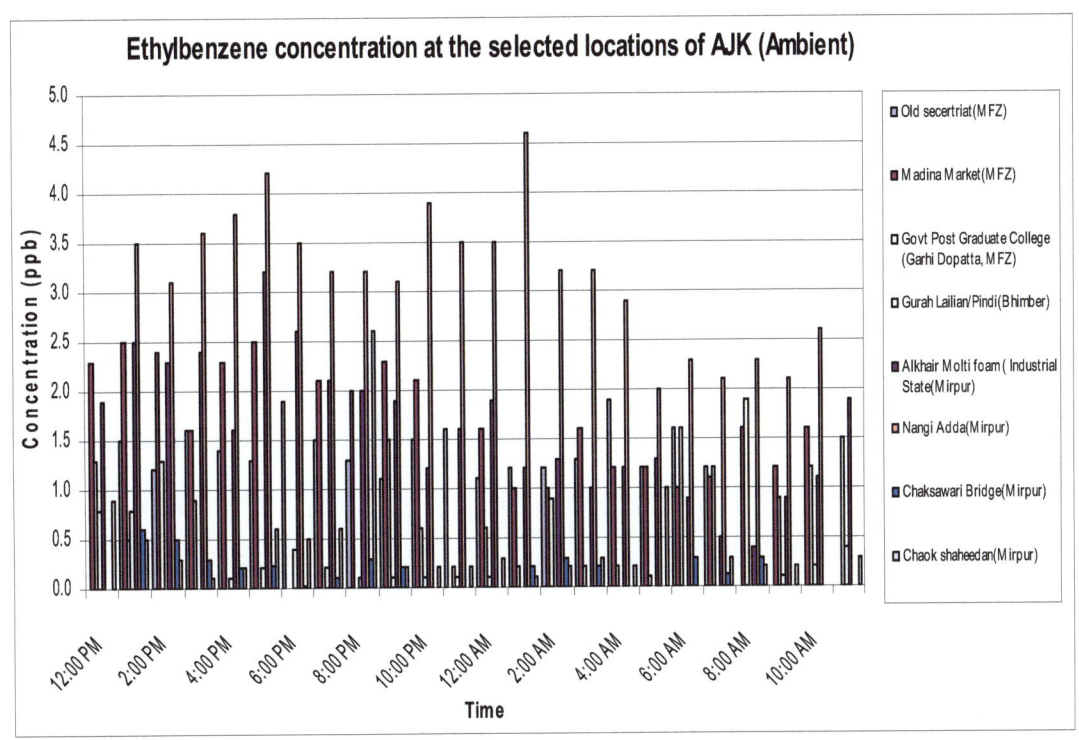

Fig 6.16

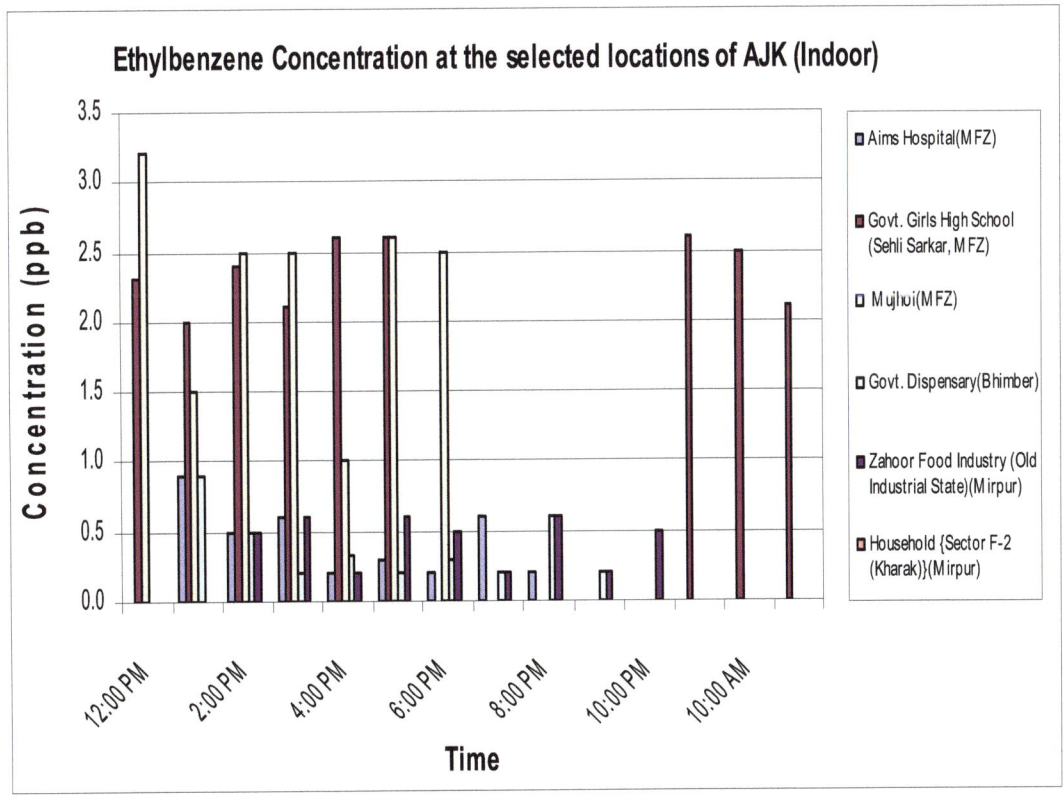

Fig 6.17

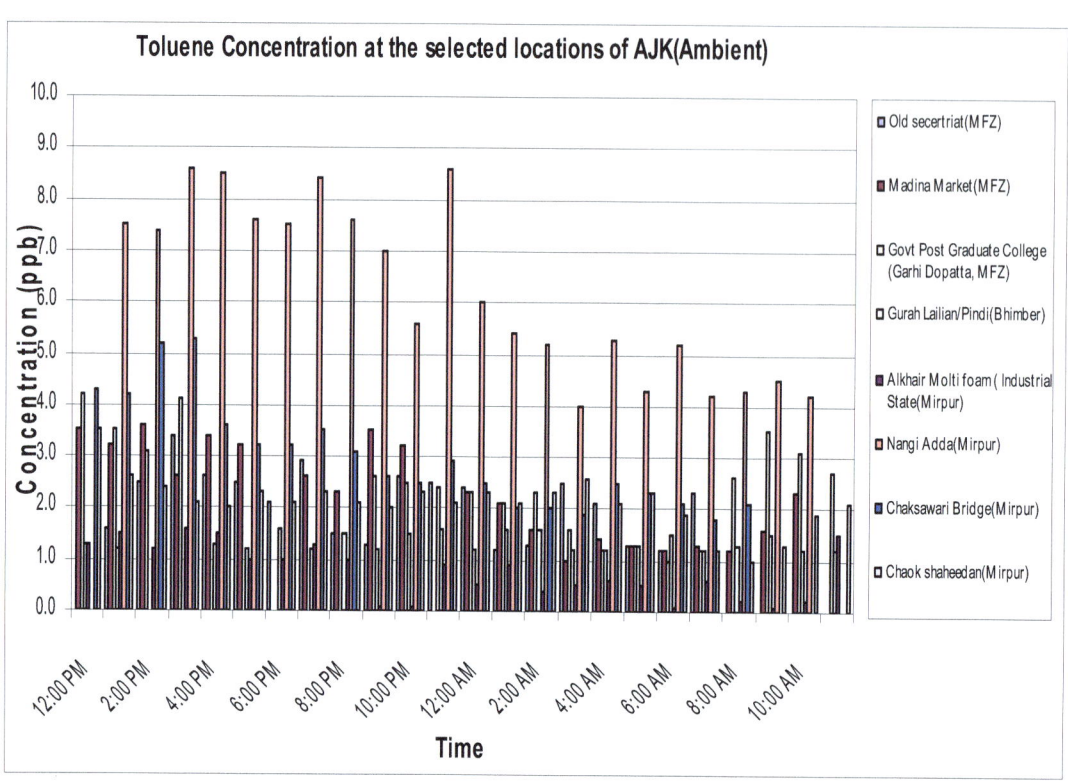

Fig 6.18

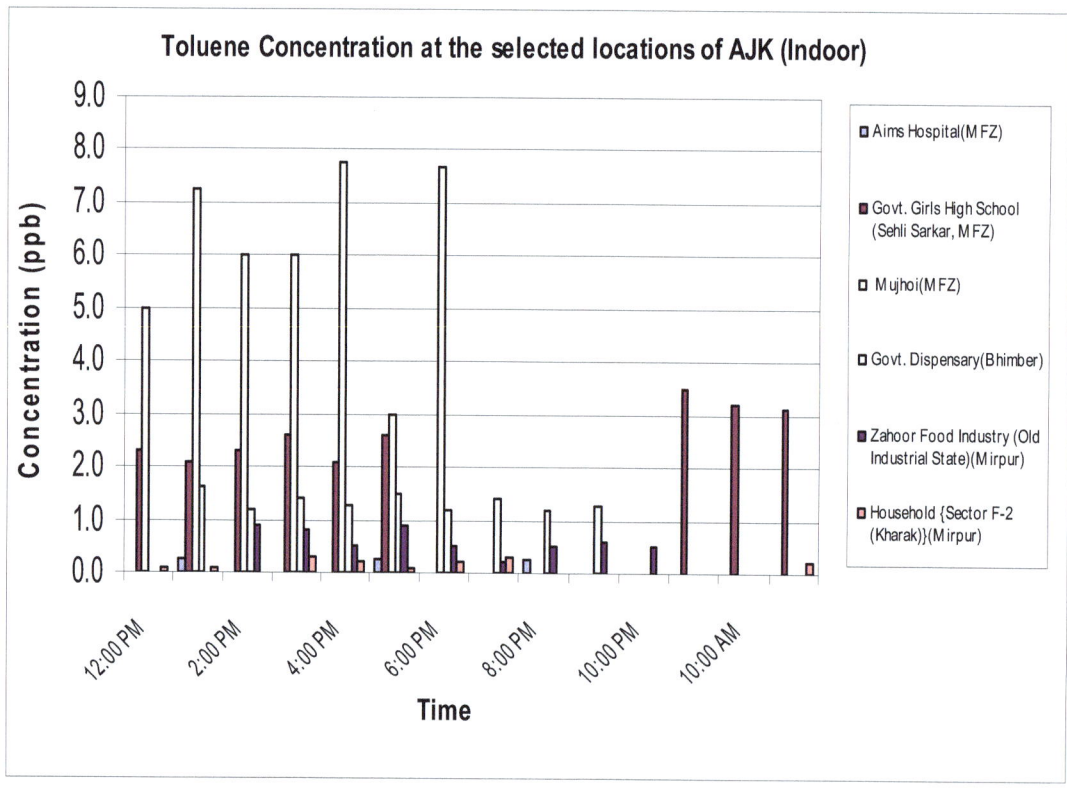

Fig 6.19

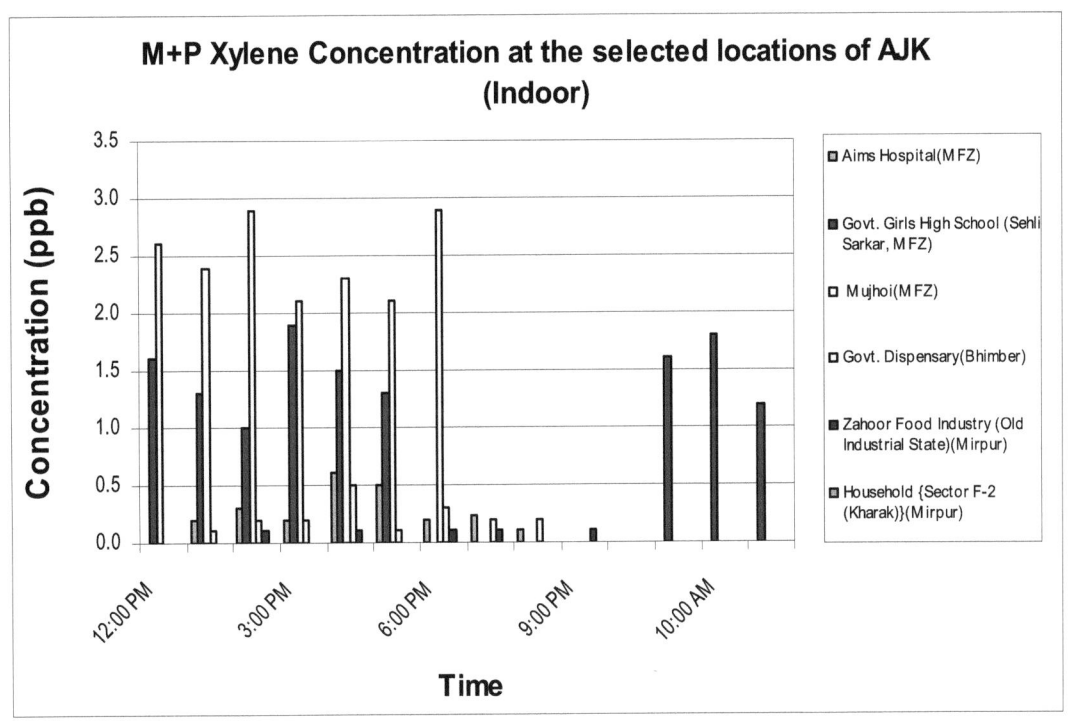

Fig 6.20

Annual average BTEX concentrations at Northbrook and Chicago-Jardine to are benzene 0.29 and 0.29 ppb, toluene 0.44 and 0.41 ppb, m/p Xylene 0.18 and 0.16 ppb, respectively. Concentrations at Schiller Park were found to be much higher than those at Northbrook or Chicago-Jardine with benzene at 0.42 ppb, toluene at 0.68 ppb and m/p Xylene at 0.31 ppb. Similar trend has been observed in different locations of Azad Jammu and Kashmir. However in some cases in AJK even lower than detection limit values has been observed and in some cases only spike of higher value is measured. Overall the observed values are lower and do not pose any risk.

6.11 Bioaerosols

Bioaerosols are biological aerosols which contain living organisms or are released from living organisms. Dust mites, fungi, spores, pollen, bacteria, viruses, amoebas, fragments of plant materials, and human and pet dander are some examples. Since bioaerosols are potentially related to various human health effects and the indoor environment provides a unique exposure situation, concerns about indoor bioaerosols have increased over the last decade. Despite its importance and complexity, the indoor environment has been much less thoroughly been studied in Pakistan than that outdoors. Fungi and their biological metabolites, the *mycotoxins*, are present almost everywhere in indoor and outdoor environments. The most common symptoms of these silent and ruthless attackers of human health are runny nose, eye irritation, cough, congestion, and aggravation of asthma.

These biological chemicals can arise from a host of means, but there are two common classes: (a) moisture induced growth of mold colonies and (b) natural substances released into the air such as animal dander and plant pollen. Moisture buildup inside buildings may arise from water penetrating compromised areas of the building envelope or skin, from plumbing leaks, from condensation due to improper ventilation, or from ground moisture penetrating a building part. In areas where cellulosic materials (paper and wood, including drywall) become moist and fail to dry within 48 hours, mold mildew can propagate and release allergenic spores into the air.

In many cases, if materials have failed to dry out several days after the suspected water event, mold growth is suspected within wall cavities even if it is not immediately visible. Through a mold investigation, which may include destructive inspection, one should be able to determine the presence or absence of mold. In a situation where there is visible mold and the indoor air quality may have been compromised, mold remediation may be needed. Mold testing and inspections should be done by an independent investigator to avoid any conflict of interest and to insure accurate results; free mold testing offered by remediation companies is not recommended.

There are some varieties of mold that contain toxic compounds (mycotoxins). However, exposure to hazardous levels of mycotoxin via inhalation is not possible in most cases, as toxins are produced by the fungal body and are not at significant levels in the released spores. The primary hazard of mold growth, as it relates to indoor air quality, comes from the allergenic properties of the spore cell wall. More serious than most allergenic properties is the ability of mold to trigger episodes in persons that already have asthma, a serious respiratory disease.

Mold is always associated with moisture and its growth can be inhibited by keeping humidity levels below 50%. Moisture problems causing Mold growth can be direct such as a water leaks and/or indirect such as condensation due to humidity levels.

Bioaerosols Concentrations (CFU/m^3) at various location of AJK

SAMPLING SITE	TOTAL MICROBIAL COUNT	AIR MOLD COUNT SPECIES	CFUs	Volume of Air	Air Concentration CFU/m^3
Govt Girls Higher Secondary School Muzaffarabad	> 100	TOTAL	6	0.1415	42.403
		ALTERINARIA SPECIES	2	0.1415	14.134
		CLADOSPORIUM SPECIES	4	0.1415	28.269
AIMS Hospital Muzaffarabad	> 100	TOTAL	17	0.1415	120.141
		PENICILLIUM SPECIES	6	0.1415	42.403
		DRECHSLERA	4	0.1415	28.269
		ASPERGILLUS FLAVUS	2	0.1415	14.134
		CLADOSPORIUM SPECIES	5	0.1415	35.336
Mirpur Industrial Area	> 100	TOTAL	34	0.1415	240.283
		ASPERGILLUS FLAVUS	23	0.1415	162.544
		ASPERGILLUS NIGER	11	0.1415	77.739
Garhi Dupatta Muzaffarabad	> 100	TOTAL	31	0.1415	219.081
		MALBRANCHING SPECIES	18	0.1415	127.208
		ULOCLADIUM SPECIES	1	0.1415	7.067
		EXOPHILA SPECIES	4	0.1415	28.269
		PENICILLIUM SPECIES	1	0.1415	7.067
		BIPOLARIUS SPECIES	7	0.1415	49.470
Bhimber	> 100	TOTAL	5	0.1415	35.336
		DRECHSLER	1	0.1415	7.067
		ALTERINARIA SPECIES	3	0.1415	21.201
		EXOPHILA SPECIES	1	0.1415	7.067
Mirpur house Hold	> 100	TOTAL	3	0.1415	21.201
		CLADOSPORIUM SPECIES	1	0.1415	7.067
		RHIZOPUS	1	0.1415	7.067
		ALTERINARIA SPECIES	1	0.1415	7.067

In this study characterization and counts of molds in six indoor locations in Azad Jammu & Kashmir AJ&K were performed. These locations included household, hospital, school and industry. The indoor air was impacted for 5 minutes through single stage impactor at air flow rate 28.3 liter per minute, onto fungi specific agar medium contained in a 100 mm Petri dish fitted under the sieve. Occupants were allowed to use the common areas normally during sampling. The number of colony-forming units (CFUs) per cubic meter was calculated from the number of CFUs counted per plate and the collected air volumes.

The indoor fungal levels were varied from 21 to 240 CFUs per cubic meter with predominant genera of *Penicillium, Aspergillus Flavus, Cladosporium and Malbranching* in all samples. Since there are as such no laws governing indoor fungal bioaerosols levels, fungal loading levels were compared against currently available guidelines by American Conference of Governmental Industrial Hygienists (ACGIH) and US Public Health Services. Using these guidelines, the results indicate that fungal organisms' at most monitoring sites meet the recommended level of 200 CFUs per cubic meter except two sites i.e. rural household and food industry with slightly at higher side.

Colbeck, Nasir et al (2008) examined 42 houses in urban and rural areas of Pakistan for Bioaerosols. The air samples were taken with an Anderson sixstage viable particle sampler, loaded with Malt Extract Agar, MacConkey Agar, and Trypticase Soy Agar. In Lahore, the highest total bacteria (13,900 CFU/m^3) and fungal (5,300 CFU/m^3) concentrations were found among houses in slums. However, the outdoor levels were generally higher than those indoors. The highest outdoor concentration of total bacteria and fungi was 20,700 and 3,300 CFU/m^3, respectively. On the other hand, in rural sites, the maximum concentration of total bacteria and fungi was 29,200 and 32,800 CFU/m^3. The indoor levels of bioaerosols were higher than those outdoors in all of the samples, probably due to indoor cattle sheds and excessive use of wood as construction materials.

CHAPTER SEVEN: CONCLUSIONS

The Data shows that the levels of pollution of NO_x and SO_2 are within the US EPA limits. All the values of NO_x recorded at different location in AJK are within the US EPA, WHO (200 µg/m^3) and NEQS (80+40 µg/m^3)values Average values of SO_2 all the sites are measured in AJK are within the US EPA, World Bank and NEQS values of 80µg/m^3 (annual) and 120µg.m^{-3} (24hrs), respectively. Overall CO level in AJK is lower than the 10 mg.m^{-3} of the NEQS values. Noise Levels measured were within the NEQS limits. Ozone concentrations measured were also within the limits. PM10 and PM$_{2.5}$ are well within the NEQS limits and were also lower than the studies conducted in different parts of Pakistan. VOCs measured in this study were also on the lower side with only a few spike values. Bioaerosols were also on a very lower side as compared to the studies conducted in other parts of the country.

CHAPTER EIGHT:
HOURLY AVERAGE FIELD MEASUREMENT DATA

Site No: One (Muzaffarabad)
Site ID: Old secretariat
Sampling Type : Ambient Air
Date: 25-03-2010

TIME	SO$_2$ (µg.m^{-3})	NO$_x$ (µg.m^{-3})	CO (mg.m^{-3})	CO$_2$ (mg.m^{-3})	O$_3$ (µg.m^{-3})	Noise (dB)	Benzene (ppb)	Toluene (ppb)	Ethylbenzene (ppb)	Xylene (ppb)	M+P Xylene (ppb)	Wind speed	Pressure	Direction	Air Temp.	Humidity
13:00	55.7	85.0	8.4	686	74.7	73.6	5.3	1.6	1.5	1.6	1.5	0.0	947	188	32.5	18.3
14:00	62.9	86.7	7.1	687	75.0	75.0	3.1	2.5	1.2	2.5	3.4	0.0	942	195	32.8	16.3
15:00	70.1	77.5	7.8	639	59.5	75.0	4.5	3.4	1.6	2.4	1.2	0.0	939	260	32.4	15.3
16:00	94.7	82.5	8.4	639	52.3	74.9	4.2	2.6	1.4	1.3	1.3	0.0	937	173	31.0	18.0
17:00	95.2	78.8	8.5	662	61.1	75.0	6.5	2.5	1.3	2.6	1.2	0.1	938	162	29.4	22.6
18:00	89.6	66.9	8.8	693	59.0	75.3	5.3	2.1	1.9	2.5	1.2	0.0	941	200	26.7	33.6
19:00	84.6	58.7	7.5	731	54.3	77.3	4.3	2.9	1.5	2.3	1.9	0.1	944	160	25.0	40.1
20:00	81.7	62.5	4.9	749	47.7	75.5	5.1	1.5	1.3	2.4	1.5	0.0	948	140	23.4	46.7
21:00	77.3	56.2	4.8	711	40.0	74.8	3.2	1.3	1.1	2.3	2.6	0.1	949	122	21.7	52.3
22:00	74.1	60.0	4.7	621	26.7	70.7	7.8	2.6	1.5	2.6	2.4	0.0	951	144	21.1	53.7
23:00	63.4	68.2	3.7	634	20.9	62.9	3.2	2.5	1.6	1.3	2.3	0.1	951	127	20.4	56.6
0:00	55.5	67.4	4.1	614	14.7	56.8	8.4	2.4	1.1	1.5	2.1	0.0	950	141	19.7	56.5
1:00	53.2	63.9	3.0	585	15.6	49.5	7.3	1.2	1.2	1.8	1.6	0.0	950	128	19.6	58.2
2:00	53.7	53.8	2.9	587	15.1	43.4	6.2	1.3	1.2	4.2	1.2	0.0	950	177	19.5	58.3
3:00	51.1	44.8	2.4	578	11.4	41.4	5.3	2.5	1.3	1.6	1.3	0.0	947	136	19.0	60.7
4:00	47.7	44.3	3.0	585	11.7	43.8	4.6	2.1	1.9	2.3	1.8	0.1	947	186	18.2	63.6
5:00	45.9	44.2	2.4	587	12.2	44.2	3.2	1.3	1.2	2.6	1.2	0.1	948	193	17.6	67.1
6:00	42.4	43.0	2.6	583	11.8	44.1	1.6	1.2	1.6	2.0	1.6	0.0	950	121	17.3	67.7
7:00	53.3	40.1	3.9	614	17.1	42.9	3.5	2.3	1.2	1.6	1.9	0.1	952	151	17.3	67.5
Min	42.4	40.1	2.4	578	11.4	41.4	1.6	1.2	1.1	1.3	1.2	0.0	937.3	120.5	17.3	15.3
Max	95.2	86.7	8.8	749	75.0	77.3	8.4	3.4	1.9	4.2	3.4	0.1	952.3	260.4	32.8	67.7
Average	65.9	62.3	5.2	641	35.8	61.9	4.9	2.1	1.4	2.2	1.7	0.0	946.3	163.4	23.4	46.0

Site No: Two (Muzaffarabad)
Site ID: Madina Market
Sampling Type : Ambient Air
Date:26-03-2010

TIME	SO_2 ($\mu g.m^{-3}$)	NO_x ($\mu g.m^{-3}$)	CO ($mg.m^{-3}$)	CO_2 ($mg.m^{-3}$)	O_3 ($\mu g.m^{-3}$)	Noise (dB)	Benzene (ppb)	Toluene (ppb)	Ethylbenzene (ppb)	Xylene (ppb)	M+P Xylene (ppb)	Wind speed	Pressure	Direction	Air Temp.	Humidity
12:00	55.8	56.8	7.1	638	65.8	57.9	8.6	3.5	2.3	1.2	1.6	0.2	937.4	194.1	27.9	26.6
13:00	56.6	57.3	5.7	681	49.6	55.9	8.4	3.2	2.5	1.3	1.2	0.1	933.7	135.8	28.7	26.5
14:00	61.6	58.7	6.2	622	55.9	74.9	7.5	3.6	2.4	1.5	1.2	0.6	931.7	198.1	26.3	32.6
15:00	56.1	55.5	7.0	575	71.7	68.8	8.4	2.6	1.6	1.0	1.9	0.0	929.5	168.9	24.1	49.4
16:00	55.8	53.8	6.0	572	42.1	73.3	8.3	3.4	2.3	1.2	0.9	0.2	931.2	146.1	22.9	50.7
17:00	53.2	53.4	7.2	604	35.8	75.9	8.1	3.2	2.5	1.6	0.5	1.3	933.1	239.2	21.5	51.8
19:00	52.7	53.2	6.0	585	79.5	56.8	7.3	2.6	2.1	1.2	0.6	1.8	935.6	287.2	21.9	40.4
20:00	51.1	44.2	6.1	576	69.4	50.0	8.6	2.3	2.0	1.0	0.5	0.9	947.3	182.0	20.8	48.4
21:00	50.8	43.5	3.8	584	44.9	47.8	8.5	3.5	2.3	1.2	0.2	0.3	959.2	143.0	17.2	75.9
22:00	48.7	38.2	3.8	575	31.1	50.2	7.2	3.2	2.1	0.8	0.6	0.2	963.1	123.4	16.0	88.3
0:00	48.5	37.8	3.0	567	19.9	43.6	6.5	2.3	1.6	0.5	0.5	0.2	956.2	165.7	15.6	89.3
1:00	43.2	38.2	2.7	565	20.9	46.5	5.3	2.1	1.0	0.3	0.2	0.0	952.3	238.6	15.9	82.7
2:00	47.9	36.7	2.9	565	21.8	45.0	4.3	1.6	1.0	0.2	0.9	0.0	950.3	115.1	15.8	81.2
3:00	39.8	34.4	2.6	574	21.0	47.6	5.1	1.0	1.6	0.3	0.5	0.0	949.5	153.8	15.3	84.3
4:00	37.2	34.6	2.5	573	17.0	49.4	4.0	1.4	1.2	0.2	0.2	0.0	950.3	172.9	14.9	86.1
5:00	39.8	31.0	2.6	574	16.5	48.9	3.5	1.3	1.2	0.3	0.3	0.0	949.7	148.1	14.5	86.4
6:00	37.2	34.4	2.7	569	15.4	46.5	3.2	1.2	1.0	0.2	0.1	0.0	950.3	154.9	14.1	88.6
7:00	27.0	32.3	2.7	569	14.7	42.3	4.5	1.3	1.1	0.3	0.2	0.0	950.2	148.7	13.8	88.2
8:00	35.4	28.8	3.0	572	22.8	41.2	2.3	1.2	1.6	0.2	0.3	0.0	952.5	152.4	13.8	88.3
9:00	45.1	38.7	3.7	580	42.5	45.3	3.5	1.6	1.2	0.6	1.5	0.1	953.6	173.6	14.4	86.1
10:00	48.5	45.7	3.7	585	49.9	42.1	3.6	2.3	1.6	1.2	1.2	0.1	955.0	143.3	17.5	72.0
Min	27.0	28.8	2.5	565	14.7	41.2	2.3	1.0	1.0	0.2	0.1	0.0	929.5	115.1	13.8	26.5
Max	61.6	58.7	7.2	681	79.5	75.9	8.6	3.6	2.5	1.6	1.9	1.8	963.1	287.2	28.7	89.3
Average	47.2	43.2	4.3	586	38.5	52.8	6.0	2.3	1.7	0.8	0.7	0.3	946.3	170.7	18.7	67.8

Site No: Three(Muzaffarabad)
Site ID: Aims Hospital
Sampling Type : Indoor Air
Date:27-03-2010

TIME	SO_2 ($\mu g.m^{-3}$)	NO_x ($\mu g.m^{-3}$)	CO ($mg.m^{-3}$)	CO_2 ($mg.m^{-3}$)	O_3 ($\mu g.m^{-3}$)	Noise (dB)	Benzene (ppb)	Toluene (ppb)	Ethylbenzene (ppb)	Xylene (ppb)	M+P Xylene (ppb)
13:00	40.9	43.5	10.5	591	238.9	71.6	2.0	0.3	0.9	0.5	0.2
14:00	36.0	45.2	10.0	668	171.9	70.5	0.8	0.0	0.5	0.2	0.3
15:00	38.4	48.5	11.5	623	183.6	69.4	0.8	0.0	0.6	0.3	0.2
16:00	43.8	48.8	10.9	598	215.7	71.2	1.3	0.0	0.2	0.2	0.6
17:00	39.8	46.2	9.9	662	155.4	62.4	1.3	0.3	0.3	0.6	0.5
18:00	36.1	46.0	11.9	631	163.8	72.0	0.3	0.0	0.2	0.2	0.2
19:00	36.9	46.4	11.0	593	169.4	67.7	0.3	0.0	0.6	0.2	0.2
20:00	41.1	44.4	12.7	608	246.6	66.8	0.5	0.3	0.2	0.3	0.1
Min	36.0	43.5	9.9	591	155.4	62.4	0.3	0.0	0.2	0.2	0.1
Max	43.8	48.8	12.7	668	246.6	72.0	2.0	0.3	0.9	0.6	0.6
Average	39.1	46.1	11.1	622	193.2	68.9	0.9	0.1	0.4	0.3	0.3

Site No: Four (Muzaffarabad)   Sampling Type : Indoor Air

Site ID: Govt Girls High School (Sehli Sarkar)   Date: 29-03-2010

TIME	SO_2 ($\mu g.m^{-3}$)	NO_x ($\mu g.m^{-3}$)	CO ($mg.m^{-3}$)	CO_2 ($mg.m^{-3}$)	O_3 ($\mu g.m^{-3}$)	Noise (dB)	Benzene (ppb)	Toluene (ppb)	Ethyl-benzene (ppb)	Xylene (ppb)	M+P Xylene (ppb)
9:00	33.2	38.2	8.6	716	29.7	52.8	8.5	3.5	2.6	3.5	1.6
10:00	40.9	40.4	8.3	702	37.1	55.0	7.9	3.2	2.5	3.2	1.8
11:00	49.6	40.6	8.2	673	32.0	53.5	6.5	3.1	2.1	3.6	1.2
12:00	51.1	45.5	8.2	670	28.5	52.0	8.4	2.3	2.3	3.2	1.6
13:00	40.1	47.6	8.2	671	29.7	48.5	7.6	2.1	2.0	3.1	1.3
14:00	37.2	49.3	8.2	669	35.7	45.3	6.5	2.3	2.4	2.3	1.0
15:00	34.6	40.0	7.9	652	39.8	54.3	6.5	2.6	2.1	3.2	1.9
16:00	31.7	34.8	7.9	623	35.2	53.3	6.1	2.1	2.6	3.6	1.5
17:00	33.0	28.6	7.8	614	36.8	45.0	5.2	2.6	2.6	3.1	1.3
Min	31.7	28.6	7.8	614	28.5	45.0	5.2	2.1	2.0	2.3	1.0
Max	51.1	49.3	8.6	716	39.8	55.0	8.5	3.5	2.6	3.6	1.9
Average	39.0	40.5	8.2	666	33.8	51.1	7.0	2.6	2.4	3.2	1.5

Site No: Five
Sampling Type : Indoor Air
Site ID: Mujhoi
Date:30-03-2010

TIME	SO_2 ($\mu g.m^{-3}$)	NO_x ($\mu g.m^{-3}$)	CO ($mg.m^{-3}$)	CO_2 ($mg.m^{-3}$)	O_3 ($\mu g.m^{-3}$)	Noise (dB)	Benzene (ppb)	Toluene (ppb)	Ethylbenzene (ppb)	Xylene (ppb)	M+P Xylene (ppb)
12:00	42.4	105.8	5.6	746	39.2	46.3	12.0	5.0	3.2	3.5	2.6
13:00	42.3	68.3	6.1	681	38.8	47.5	14.0	7.3	1.5	4.2	2.4
14:00	37.8	81.1	5.4	698	36.1	46.3	12.8	6.0	2.5	3.6	2.9
15:00	31.4	60.0	5.3	689	33.8	47.3	11.8	6.0	2.5	2.5	2.1
16:00	33.7	66.7	5.6	702	31.7	44.5	11.8	7.8	1.0	2.9	2.3
17:00	38.5	87.1	6.1	681	29.9	44.8	11.8	3.0	2.6	3.5	2.1
18:00	40.4	77.6	5.6	636	29.6	43.7	11.3	7.7	2.5	2.9	2.9
Min	31.4	60.0	5.3	636	29.6	43.7	11.3	3.0	1.0	2.5	2.1
Max	42.4	105.8	6.1	746	39.2	47.5	14.0	7.8	3.2	4.2	2.9
Average	38.1	78.1	5.7	690	34.2	45.7	12.2	6.1	2.3	3.3	2.5

	Site No: Six(Muzaffarabad)												Sampling Type : Ambient Air			
	Site ID: Govt Post Graduate College												Date:30-03-2010			
TIME	SO_2 ($\mu g.m^{-3}$)	NO_x ($\mu g.m^{-3}$)	CO ($mg.m^{-3}$)	CO_2 ($mg.m^{-3}$)	O_3 ($\mu g.m^{-3}$)	Noise (dB)	Benzene(ppb)	Toluene (ppb)	Ethyl benzene (ppb)	Xylene (ppb)	M+P Xylene (ppb)	Wind speed	Pressure	Direction	Air Temp	Humidity
21:00	39.2	49.6	3.8	585	23.2	44.8	6.5	2.6	1.5	0.5	1.6	2.0	900.3	163	17.1	64.1
22:00	40.0	52.6	3.8	585	22.5	44.8	3.5	2.5	0.6	0.6	1.2	1.3	895.5	153	16.9	58.3
23:00	40.2	49.4	4.1	586	25.2	49.5	4.2	2.4	0.2	0.1	1.5	0.3	899.0	193	16.4	65.1
0:00	40.9	49.8	3.9	588	24.2	47.0	3.6	2.3	0.6	0.9	1.9	0.1	900.6	185	14.3	80.2
1:00	37.2	48.6	3.9	585	22.0	53.8	2.3	2.1	0.2	0.8	1.2	0.1	900.9	250	13.1	87.2
2:00	34.6	52.4	4.0	578	21.1	53.5	2.1	2.3	0.9	0.2	1.3	0.0	901.6	224	12.5	90.7
3:00	36.7	48.6	4.1	601	21.3	50.0	2.3	1.6	0.2	0.6	0.9	0.0	902.3	232	11.9	92.5
4:00	33.0	47.0	4.1	593	21.1	47.3	1.2	1.2	0.2	0.2	0.5	0.0	902.0	195	11.4	93.5
5:00	31.4	49.0	3.8	653	18.7	48.8	1.5	1.3	0.1	0.3	0.4	0.0	902.2	202	10.9	94.3
6:00	35.4	50.7	4.1	620	21.2	45.5	1.0	1.0	1.6	0.2	0.3	0.0	901.9	158	10.6	94.8
7:00	39.2	52.4	3.7	599	19.5	44.8	1.2	1.2	1.2	0.5	0.2	0.0	901.6	172	10.3	95.1
8:00	38.0	51.8	3.7	604	23.1	44.5	1.6	2.6	1.9	2.3	0.1	0.0	902.4	181	10.0	95.5
9:00	41.8	46.9	4.1	613	36.9	48.5	2.1	3.5	0.9	0.2	0.2	0.0	903.4	132	9.7	95.7
10:00	40.1	50.9	4.1	621	46.2	50.8	2.6	3.1	1.2	0.1	0.3	0.0	905.6	199	10.2	96.2
11:00	41.3	51.5	4.4	617	55.4	52.8	3.5	2.7	1.5	0.3	1.0	0.1	905.8	142	16.1	86.4
12:00	43.4	48.6	4.7	593	61.8	52.8	3.5	4.2	1.3	0.2	1.2	0.2	906.1	153	17.5	79.3
13:00	45.7	49.6	4.7	621	64.4	48.8	4.6	3.5	0.5	0.2	1.6	0.6	907.2	158	19.2	71.2
14:00	54.0	58.7	4.5	593	65.6	47.8	7.1	3.1	1.3	0.3	1.2	0.1	906.5	174	24.3	68.4
15:00	62.4	63.5	3.8	610	67.4	46.3	6.5	4.1	0.9	0.2	1.3	0.5	907.4	168	26.5	62.1
Min	31.4	46.9	3.7	578	18.7	44.5	1.0	1.0	0.1	0.1	0.1	0.0	895.5	132.4	9.7	58.3
Max	62.4	63.5	4.7	653	67.4	53.8	7.1	4.2	1.9	2.3	1.9	2.0	907.4	250.3	26.5	96.2
Average	40.8	51.1	4.1	602	34.8	48.5	3.2	2.5	0.9	0.5	0.9	0.3	902.8	180.8	14.7	82.7

Site No: One (Bhimber)　　　　　　　　　　　　　　　Sampling Type : Ambient Air
Site ID: Gurah Lailian/Pindi　　　　　　　　　　　　　Date:01-04-2010

TIME	SO_2 ($\mu g.m^{-3}$)	NO_x ($\mu g.m^{-3}$)	CO ($mg.m^{-3}$)	CO_2 ($mg.m^{-3}$)	O_3 ($\mu g.m^{-3}$)	Noise (dB)	Benzene (ppb)	Toluene (ppb)	Ethyl Benzene (ppb)	Xylene (ppb)	M+P Xylene (ppb)	Wind speed	Press.	Dir.	Air Temp	Humidity
16:00	60.3	61.1	4.5	621	109.4	62.0	9.5	1.3	0.1	0.0	0.1	2.1	1007.3	185	35.1	49.1
17:00	64.5	60.5	4.8	652	111.3	63.3	8.6	1.2	0.2	0.1	0.0	1.6	1006.8	145	34.9	51.3
18:00	60.3	64.1	4.1	693	111.7	61.8	7.3	1.6	0.4	0.0	0.1	1.2	1008.3	196	34.1	53.9
19:00	62.9	66.2	3.7	675	106.6	61.8	8.3	1.2	0.2	0.1	0.0	1.5	1007.8	178	33.4	57.2
20:00	65.2	68.1	4.7	716	105.1	63.5	8.2	1.5	0.1	0.0	0.0	0.9	1008.2	152	31.2	59.2
21:00	65.5	64.1	6.0	716	97.6	62.0	8.3	1.2	0.1	0.2	0.0	0.5	1009.1	145	30.6	60.2
22:00	62.9	61.3	4.1	713	107.0	61.0	7.2	1.5	0.1	0.0	0.0	0.2	1007.0	163	29.4	60.8
23:00	59.5	66.2	3.7	693	85.8	61.5	9.6	1.6	0.1	0.1	0.0	0.3	1007.1	201	28.6	60.8
0:00	62.9	58.7	4.1	657	79.6	62.3	8.0	1.2	0.1	0.0	0.0	0.0	1005.2	152	27.3	61.2
1:00	59.5	68.1	4.8	711	72.8	64.5	8.2	1.6	0.0	0.0	0.0	0.0	1003.3	163	27.1	61.8
2:00	59.2	66.0	4.0	621	66.8	61.5	7.6	1.6	0.0	0.0	0.0	0.0	1003.2	145	26.5	62.4
3:00	64.7	61.3	3.6	652	64.8	62.8	8.5	1.2	0.0	0.0	0.0	0.0	1001.3	185	26.0	63.5
4:00	59.2	64.3	4.1	634	59.7	60.8	7.2	1.2	0.0	0.0	0.0	0.0	999.3	196	25.9	63.9
5:00	65.5	58.7	3.7	585	56.3	58.3	7.6	1.3	0.0	0.0	0.0	0.0	999.2	178	25.1	58.3
6:00	64.5	68.1	3.7	637	79.0	63.0	8.2	1.5	0.0	0.0	0.0	0.2	997.3	201	24.6	52.1
7:00	59.5	58.7	4.1	652	93.9	63.0	7.2	1.2	0.0	0.0	0.0	0.0	993.2	206	26.3	49.3
8:00	70.0	66.2	4.0	662	100.6	67.0	8.3	1.3	0.0	0.0	0.0	0.0	996.2	152	28.1	46.0
9:00	63.1	68.1	3.7	652	126.3	65.5	8.2	1.5	0.1	0.0	0.0	0.0	1000.6	136	28.9	43.9
10:00	65.5	66.2	3.6	646	109.8	66.3	6.0	1.2	0.2	0.1	0.1	0.0	1003.6	120	30.2	36.2
11:00	74.9	66.0	4.8	634	107.9	64.5	7.2	1.2	0.4	0.2	0.3	0.0	1005.4	145	31.6	32.7
12:00	64.5	68.1	4.0	637	128.2	65.0	5.2	1.3	0.8	0.1	0.2	0.9	1007.2	163	32.4	29.5
13:00	59.2	71.6	3.7	664	120.7	64.0	4.1	1.2	0.8	0.2	0.1	0.2	1008.1	152	34.2	25.1
Min	59.2	58.7	3.6	585	56.3	58.3	4.1	1.2	0.0	0.0	0.0	0.3	993.2	120.0	24.6	25.1
Max	74.9	71.6	6.0	716	128.2	67.0	9.6	1.6	0.8	0.2	0.3	0.0	1009.1	206.0	35.1	63.9
Average	63.3	64.6	4.2	660	95.5	63.0	7.7	1.3	0.2	0.1	0.0	2.1	1003.9	166.3	29.6	51.7

Site No: Two (Bhimber) — Site ID: Govt. Dispensary — Sampling Type: Indoor Air — Date: 02-04-2010

TIME	SO_2 (µg.m^{-3})	NO_x (µg.m^{-3})	CO (mg.m^{-3})	CO_2 (mg.m^{-3})	O_3 (µg.m^{-3})	Noise (dB)	Benzene (ppb)	Toluene (ppb)	Ethylbenzene (ppb)	Xylene (ppb)	M+P Xylene (ppb)
15:00	28.3	38.7	9.1	684	90.5	49.3	5.2	1.6	0.9	0.2	0.1
16:00	27.8	38.7	9.7	695	90.7	47.0	4.6	1.2	0.5	0.5	0.2
17:00	25.7	40.4	9.8	798	82.8	46.5	3.5	1.4	0.2	0.3	0.2
18:00	26.5	40.0	8.7	877	74.5	45.3	3.2	1.3	0.3	0.2	0.5
19:00	28.0	45.7	9.7	802	75.8	46.5	4.0	1.5	0.2	0.3	0.1
20:00	31.7	39.9	9.9	678	64.6	49.0	3.1	1.2	0.3	0.1	0.3
21:00	27.2	40.6	10.4	690	65.5	50.0	2.3	1.4	0.2	0.2	0.2
22:00	28.8	39.9	9.0	684	70.2	48.0	2.1	1.2	0.6	0.3	0.2
23:00	27.8	40.6	10.9	724	67.4	44.0	2.3	1.3	0.2	0.2	0.0
Min	25.7	38.7	8.7	678	64.6	44.0	2.1	1.2	0.2	0.1	0.0
Max	31.7	45.7	10.9	877	90.7	50.0	5.2	1.6	0.9	0.5	0.5
Average	28.0	40.5	9.7	737	75.8	47.3	3.4	1.3	0.4	0.3	0.2

Site No: One (Mirpur) | | | | | | | | | | Sampling Type : Indoor Air
Site ID: Zahoor Food Industry (Old Industrial State) | | | | | | | | | | Date: 03-04-2010

TIME	SO_2 ($\mu g.m^{-3}$)	NO_x ($\mu g.m^{-3}$)	CO ($mg.m^{-3}$)	CO_2 ($mg.m^{-3}$)	O_3 ($\mu g.m^{-3}$)	Noise (dB)	Benzene (ppb)	Toluene (ppb)	Ethylbenzene (ppb)	Xylene (ppb)	M+P Xylene (ppb)
14:00	28.0	65.4	9.5	669	70.6	54.5	1.5	0.9	0.5	0.0	0.1
15:00	27.2	50.0	9.4	697	85.8	57.3	1.2	0.8	0.6	0.0	0.0
16:00	27.2	65.6	9.4	682	62.7	55.3	1.6	0.5	0.2	0.5	0.1
17:00	27.0	45.7	9.4	684	46.2	54.5	1.2	0.9	0.6	0.0	0.0
18:00	27.2	48.3	9.5	681	50.9	53.5	1.5	0.5	0.5	0.3	0.1
19:00	30.7	47.8	10.1	693	52.4	53.5	1.2	0.2	0.2	0.2	0.1
20:00	31.2	45.1	10.0	665	54.1	49.0	1.3	0.5	0.6	0.0	0.0
21:00	28.8	48.9	11.7	669	49.6	48.8	1.2	0.6	0.2	0.2	0.1
22:00	24.4	46.1	11.2	675	42.4	48.0	1.2	0.5	0.5	0.0	0.0
Min	24.4	45.1	9.4	665	42.4	48.0	1.2	0.2	0.2	0.0	0.0
Max	31.2	65.6	11.7	697	85.8	57.3	1.6	0.9	0.6	0.5	0.1
Average	28.0	51.4	10.0	679	57.2	52.7	1.3	0.6	0.4	0.1	0.1

Site Two: (Mirpur)
Site ID: Al-khair Molti foam (Industrial State)
Sampling Type : Ambient Air
Date: 05-04-2010

TIME	SO₂ (μg.m⁻³)	NOₓ (μg.m⁻³)	CO (mg.m⁻³)	CO₂ (mg.m⁻³)	O₃ (μg.m⁻³)	Noise (dB)	Benzene (ppb)	Toluene (ppb)	Ethylbenzene (ppb)	Xylene (ppb)	M+P Xylene (ppb)	Wind speed	Press.	Dir.	Air Temp	Humidity
12:00	32.5	46.1	4.1	812	143.4	66.5	6.9	1.3	1.9	0.9	1.2	1.3	1014.3	205	32.6	35.2
13:00	33.0	49.3	3.7	709	114.5	71.3	5.6	1.5	2.5	0.5	1.2	1.6	1015.3	153	34.8	30.2
14:00	35.4	47.2	4.1	707	108.7	65.5	5.4	1.2	2.3	0.3	0.9	2.1	1016.2	204	35.0	28.6
15:00	35.6	40.6	4.0	716	113.2	64.5	5.6	1.6	2.4	0.2	0.5	1.5	1015.6	206	35.6	27.2
16:00	34.8	47.6	4.9	771	126.7	65.3	5.2	1.5	1.6	0.2	0.6	1.2	1015.1	196	35.4	24.3
17:00	34.6	45.5	3.7	600	125.8	63.3	5.4	1.0	3.2	0.3	0.4	1.3	1014.7	187	34.9	28.9
18:00	34.8	49.3	4.1	593	120.9	61.0	4.6	1.0	2.6	0.2	1.0	1.6	1014.1	163	34.1	31.6
19:00	30.1	47.2	3.6	613	121.8	59.8	5.3	1.3	2.1	0.3	1.2	1.8	1013.8	245	33.6	35.2
20:00	26.7	40.0	4.0	632	94.8	58.5	5.1	1.0	2.0	0.2	0.8	2.1	1013.1	235	32.8	38.6
21:00	26.7	37.2	3.7	652	98.4	53.5	5.2	0.1	1.9	0.1	0.5	1.3	1012.2	145	32.2	46.2
22:00	27.5	34.4	3.6	636	102.9	52.0	4.9	0.1	1.2	0.2	0.4	1.2	1011.2	163	31.9	53.1
23:00	28.0	34.0	3.3	671	95.0	48.8	4.3	0.9	1.6	0.5	0.6	2.6	1010.3	285	31.4	54.2
0:00	27.8	31.0	3.2	665	96.3	48.8	3.2	0.5	1.9	0.3	0.2	2.9	1009.3	254	30.5	58.2
1:00	25.9	28.0	2.9	695	92.7	51.5	3.2	0.9	1.2	0.2	0.6	3.4	1007.8	196	29.8	62.5
2:00	24.4	34.8	2.8	705	98.4	47.0	3.4	0.4	1.3	0.5	0.2	3.5	1007.1	166	29.2	62.9
3:00	23.8	34.8	3.3	716	111.3	47.5	3.6	0.5	1.0	0.3	0.6	2.6	1006.3	154	28.6	63.5
4:00	23.8	33.1	3.3	691	109.4	48.0	3.5	0.6	1.2	0.2	0.6	3.9	1005.2	178	28.0	64.1
5:00	23.3	37.6	2.9	709	107.9	50.5	3.9	0.5	1.3	0.3	0.4	3.4	1003.3	201	27.3	65.3
6:00	23.1	37.2	3.0	716	101.7	47.5	3.5	0.1	0.9	0.5	1.0	3.1	1001.2	205	25.4	65.9
7:00	22.8	35.5	2.8	735	110.0	50.0	3.4	0.6	0.5	0.2	1.7	2.9	1004.2	204	24.1	60.1
8:00	24.9	38.2	3.0	703	112.1	53.5	4.6	0.2	0.4	0.0	1.0	1.6	1007.1	201	27.3	58.3
9:00	25.4	40.4	4.0	699	123.9	57.3	5.2	0.1	0.9	0.2	1.3	1.1	1009.3	186	28.6	51.2
10:00	24.9	45.7	3.7	703	113.8	53.8	4.6	0.2	1.1	0.3	1.1	1.3	1011.6	178	29.7	48.3
11:00	23.8	49.1	4.1	742	112.4	52.0	5.9	1.5	1.9	0.2	1.3	1.6	1012.2	163	30.9	41.0
Min	22.8	28.0	2.8	593	92.7	47.0	3.2	0.1	0.4	0.0	0.2	1.1	1001.2	145.0	24.1	24.3
Max	35.6	49.3	4.9	812	143.4	71.3	6.9	1.6	3.2	0.9	1.7	3.9	1016.2	285.0	35.6	65.9
Average	28.1	40.2	3.6	691	110.7	55.7	4.6	0.8	1.6	0.3	0.8	2.1	1010.4	194.7	31.0	47.3

Site No: Three(Mirpur) Sampling Type : Ambient Air
Site ID: Nangi Adda Date: 06-04-2010

TIME	SO_2 (µg.m^{-3})	NO_x (µg.m^{-3})	CO (mg.m^{-3})	CO_2 (mg.m^{-3})	O_3 (µg.m^{-3})	Noise (dB)	Benzene (ppb)	Toluene (ppb)	Ethylbenzene (ppb)	Xylene (ppb)	M+P Xylene (ppb)	Windspeed	Press.	Direction	Air Temp.	Humidity
13:00	61.8	56.4	5.2	642	116.6	76.8	10.3	7.5	3.5	2.3	1.5	3.1	1018.4	205	32.0	18
14:00	66.5	58.7	4.9	856	119.6	75.3	9.5	7.4	3.1	2.1	1.0	2.9	1018.6	196	33.0	15
15:00	66.3	66.6	3.5	957	126.5	76.5	9.4	8.6	3.6	2.0	1.2	3.5	1018.9	187	33.0	18
16:00	63.1	56.8	3.7	876	135.9	76.0	9.8	8.5	3.8	2.3	1.0	3.1	1019.4	145	32.0	14
17:00	53.2	56.0	3.6	813	133.1	74.5	8.7	7.6	4.2	2.1	1.6	2.6	1019.9	163	30.0	23
18:00	55.8	60.7	3.7	786	117.1	75.0	9.5	7.5	3.5	2.1	1.5	2.8	1019.1	205	29.0	25
19:00	66.3	58.7	4.1	886	91.2	73.8	9.4	8.4	3.2	2.5	1.0	3.5	1018.6	201	28.0	28
20:00	55.5	64.3	3.6	725	77.9	69.0	9.5	7.6	3.2	2.3	1.2	3.4	1017.8	196	25.0	32
21:00	64.5	59.4	3.7	851	50.9	73.0	9.4	7.0	3.1	2.1	1.5	2.6	1017.1	187	24.0	36
22:00	55.8	61.1	4.0	812	28.9	68.8	9.5	5.6	3.9	2.3	1.2	2.1	1016.8	145	23.0	41
23:00	56.6	59.2	3.7	762	39.6	68.8	9.7	8.6	3.5	2.6	1.6	2.3	1015.4	163	21.0	49
0:00	61.6	61.3	4.1	708	41.7	66.8	9.5	6.0	3.5	2.1	1.2	2.5	1014.8	125	21.0	37
1:00	53.2	56.6	3.7	739	39.8	60.0	9.8	5.4	4.6	2.3	1.5	2.1	1013.1	145	19.0	53
2:00	51.1	56.0	3.6	811	33.4	58.3	9.4	5.2	3.2	2.1	1.0	3.6	1011.9	185	19.0	64
3:00	48.7	47.6	3.3	896	33.2	53.5	9.5	4.0	3.2	1.9	1.2	3.9	1009.7	163	18.0	68
4:00	47.7	45.3	2.9	874	31.5	55.5	8.9	5.3	2.9	1.5	1.6	4.1	1008.4	201	17.0	68
5:00	46.1	49.4	3.3	846	35.1	55.0	9.6	4.3	2.0	2.6	1.2	5.3	1007.6	206	16.0	61
6:00	43.2	53.0	3.6	952	33.2	55.0	9.4	5.2	2.3	2.1	1.0	3.2	1005.6	230	16.0	72
7:00	37.5	56.8	3.7	772	42.4	64.5	9.5	4.2	2.1	2.3	1.2	2.6	1006.7	245	16.0	68
8:00	48.5	58.3	3.6	700	27.8	72.3	9.8	4.3	2.3	2.1	1.2	2.1	1009.2	210	30.0	59
9:00	50.8	61.3	4.5	737	33.0	72.3	9.5	4.5	2.1	1.9	1.1	3.5	1011.3	296	27.0	38
10:00	56.3	58.3	3.6	742	45.8	73.0	9.5	4.2	2.6	2.1	1.6	2.9	1012.8	168	30.0	32
Min	37.5	45.3	2.9	642	27.8	53.5	8.7	4.0	2.0	1.5	1.0	2.1	1005.6	125.0	16.0	14.0
Max	66.5	66.6	5.2	957	135.9	76.8	10.3	8.6	4.6	2.6	1.6	5.3	1019.9	296.0	33.0	72.0
Average	55.0	57.3	3.8	806	65.2	53	8.5	3.5	2.6	3.5	1.6	3.1	1014.1	189.4	24.5	41.8

	Site No: Four(Mirpur)									Sampling Type : Indoor Air		
	Site ID:Household									Date: 07-04-2010		
TIME	SO_2 ($\mu g.m^{-3}$)	NO_x ($\mu g.m^{-3}$)	CO ($mg.m^{-3}$)	CO_2 ($mg.m^{-3}$)	O_3 ($\mu g.m^{-3}$)	Noise (dB)	Benzene(ppb)	Toluene (ppb)	Ethylbenzene (ppb)	Xylene (ppb)	M+P Xylene (ppb)
11:00	32.8	47.2	10.9	740	39.4	42.0	3.5	0.2	0.0	0.0	0.0
12:00	33.5	44.6	9.9	720	34.5	46.5	3.6	0.1	0.0	0.0	0.0
13:00	26.7	65.0	9.8	678	31.2	46.0	3.4	0.1	0.0	0.0	0.0
14:00	29.6	57.5	8.5	675	27.6	45.0	3.0	0.0	0.0	0.0	0.0
12:00	24.9	53.0	9.9	671	26.1	44.8	3.1	0.3	0.0	0.0	0.0
15:00	23.3	55.1	9.2	676	27.2	46.0	2.9	0.2	0.0	0.0	0.0
16:00	35.4	54.0	10.9	671	25.7	45.3	2.1	0.1	0.0	0.0	0.0
17:00	33.0	49.8	9.1	668	29.3	46.3	2.1	0.2	0.0	0.0	0.0
19:00	28.8	50.4	9.9	688	27.8	46.8	1.3	0.3	0.0	0.0	0.0
Min	23.3	44.6	8.5	668	25.7	42.0	1.3	0.0	0.0	0.0	0.0
Max	35.4	65.0	10.9	740	39.4	46.8	3.6	0.3	0.0	0.0	0.0
Average	29.8	53.0	9.8	687	29.9	45.4	2.8	0.2	0.0	0.0	0.0

Site No: Five(Mirpur)
Site ID: Chaksawari Bridge
Sampling Type : Ambient Air
Date:08-04-2010

TIME	SO_2 (µg.m^{-3})	NO_x (µg.m^{-3})	CO (mg.m^{-3})	CO_2 (mg.m^{-3})	O_3 (µg.m^{-3})	Noise (dB)	Benzene (ppb)	Toluene (ppb)	Ethyl benzene (ppb)	Xylene (ppb)	M+P Xylene (ppb)	Wind speed	Pressure	Direction	Air Temp	Humidity
12:00	48.5	53.8	3.7	641	77.5	61.0	4.3	1.6	0.0	0.6	0.2	3.5	1009	290	34	29
13:00	43.0	48.9	4.1	634	75.3	60.3	4.2	1.2	0.6	0.2	0.0	4.6	1009	317	35	23
14:00	48.7	45.7	4.0	657	73.2	57.3	5.2	1.0	0.5	0.3	0.0	3.5	1013	238	35	15
15:00	39.3	49.8	3.9	585	75.3	60.5	5.3	1.2	0.3	0.0	0.6	3.2	1013	289	34	24
16:00	39.8	45.1	4.9	621	73.2	63.3	3.6	1.2	0.2	0.2	0.2	4.9	1013	328	32	27
17:00	37.5	47.6	3.7	713	64.6	64.3	3.2	1.3	0.2	0.0	0.3	3.2	1014	333	32	33
18:00	44.0	46.8	3.6	641	63.1	66.3	3.2	0.9	0.0	0.0	0.2	3.4	1012	331	31	39
19:00	46.6	49.3	5.3	634	61.2	66.0	3.5	0.5	0.1	0.0	0.0	4.6	1011	316	29	44
20:00	51.1	47.2	4.0	641	58.9	62.3	3.1	0.9	0.3	0.0	0.0	3.5	1013	338	27	47
21:00	48.7	50.6	3.7	621	52.0	59.3	2.6	0.5	0.2	0.0	0.3	3.2	1014	273	25	38
22:00	45.1	53.4	3.6	635	50.3	61.3	2.5	0.9	0.0	0.0	0.2	2.6	1014	253	25	47
23:00	40.1	45.7	3.3	641	45.6	58.3	2.9	0.5	0.0	0.2	0.0	3.2	1013	271	24	53
0:00	37.2	37.8	3.2	634	41.7	56.3	2.5	0.9	0.0	0.0	0.0	3.5	1007	229	24	56
1:00	40.1	36.7	2.9	637	37.7	56.8	2.0	1.3	0.2	0.0	0.3	3.6	1013	260	23	56
2:00	37.2	35.0	2.4	621	39.6	58.8	2.0	1.0	0.3	0.0	0.2	3.4	1012	253	21	52
3:00	33.0	34.8	2.6	598	35.3	60.8	1.9	1.2	0.2	0.6	0.6	2.9	1011	271	21	64
4:00	29.3	35.0	3.0	585	31.2	57.3	2.5	1.2	0.0	0.2	0.2	3.2	1010	276	21	47
5:00	34.6	36.1	2.9	587	32.5	60.8	2.3	1.2	0.0	0.3	0.3	3.5	1004	231	21	37
6:00	29.3	33.8	3.3	603	35.3	65.5	2.1	1.3	0.3	0.2	0.2	3.6	1008	229	20	24
7:00	35.6	32.3	2.9	592	41.7	63.5	1.8	1.1	0.1	0.0	0.3	3.1	1008	260	20	13
8:00	44.3	38.2	3.6	641	45.2	67.0	2.1	0.9	0.3	0.1	0.1	3.2	1008	238	21	22
Min	29.3	32.3	2.4	585	31.2	56.3	1.8	.0.5	0.0	0.0	0.0	2.6	1004.0	229.0	20.0	13.0
Max	51.1	53.8	5.3	713	77.5	67.0	5.3	1.6	0.6	0.6	0.6	4.9	1014.0	338.0	35.0	64.0
Average	40.6	43.0	3.6	627	52.9	61.3	3.0	1.0	0.2	0.1	0.2	3.5	1010.9	277.3	26.4	37.6

Site No: Six(Mirpur)
Site ID: Chaok shaheedan
Sampling Type: Ambient Air
Date: 09-04-2010

TIME	SO_2 ($\mu g.m^{-3}$)	NO_x ($\mu g.m^{-3}$)	CO ($mg.m^{-3}$)	CO_2 ($mg.m^{-3}$)	O_3 ($\mu g.m^{-3}$)	Noise (dB)	Benzene (ppb)	Toluene (ppb)	Ethyl benzene (ppb)	Xylene (ppb)	M+P Xylene (ppb)	Wind speed	Press	Dir.	Air Temp	Humidity
12:00	60.3	66.9	4.1	866	96.9	67.3	9.5	3.5	0.9	0.2	0.4	2.6	1010	325	35	15
13:00	66.3	64.3	3.7	887	98.9	74.0	8.6	2.6	0.5	0.2	0.2	2.9	1010	326	35	15
14:00	68.1	68.6	4.0	964	96.5	74.5	8.5	2.4	0.3	0.3	0.2	3.5	1009	315	35	15
15:00	63.7	61.1	3.7	919	96.9	70.0	7.7	2.1	0.1	0.2	0.2	4.2	1008	314	33	23
16:00	55.8	68.6	4.1	793	92.4	72.5	8.5	2.0	0.2	0.0	0.3	3.5	1004	326	34	14
17:00	53.2	64.3	3.9	767	89.0	74.0	6.5	2.3	0.6	0.1	0.1	3.9	1008	328	33	23
18:00	51.1	66.2	3.7	744	84.5	74.3	8.6	2.1	0.5	0.6	0.6	4.0	1008	336	31	25
19:00	55.8	58.7	4.1	730	88.2	70.5	6.4	2.3	0.6	0.2	0.2	3.2	1008	325	29	3
20:00	66.3	60.2	3.7	748	85.2	69.0	8.0	2.1	2.6	0.2	0.3	2.9	1007	324	27	39
21:00	63.4	65.8	4.0	641	76.2	68.0	5.6	2.0	0.2	0.2	0.2	4.5	1008	310	27	34
22:00	68.9	61.3	3.5	726	65.3	68.3	6.5	2.3	0.2	0.0	0.3	3.1	1008	298	25	41
23:00	63.7	56.8	3.7	711	63.1	71.8	6.5	2.1	0.2	0.1	0.2	2.6	1006	296	24	34
0:00	55.8	55.5	3.3	760	61.2	70.0	5.3	2.3	0.3	1.0	0.2	3.5	1008	278	23	47
1:00	53.2	54.5	2.9	767	52.0	69.0	4.3	2.1	0.1	0.0	0.2	3.4	1008	263	23	44
2:00	51.9	50.8	2.8	885	43.4	66.5	5.0	2.3	0.2	0.2	0.2	2.6	1007	263	20	56
3:00	48.7	47.0	3.0	734	41.7	66.0	4.6	2.6	0.3	0.1	0.3	3.5	1007	278	19	64
4:00	51.9	40.0	2.4	827	37.7	63.8	3.2	2.1	0.2	0.1	0.1	3.1	1006	298	19	51
5:00	46.1	47.6	2.9	922	32.5	63.0	3.1	2.3	1.0	0.3	3.0	1.9	1006	285	19	6
6:00	48.5	49.3	3.3	675	30.4	65.5	3.6	1.9	0.0	0.2	0.2	2.4	1006	296	19	64
7:00	52.4	47.6	3.0	635	34.7	60.8	4.6	1.2	0.3	0.3	0.3	2.3	1007	301	19	64
8:00	56.6	55.6	4.5	660	44.1	64.5	5.3	1.0	0.2	0.2	0.2	2.8	1008	296	23	53
9:00	61.6	53.6	4.0	630	50.3	68.0	6.4	1.3	0.2	0.0	0.3	2.6	1008	315	27	39
10:00	64.5	56.8	4.1	836	61.8	71.5	5.6	1.9	0.0	0.2	0.2	3.4	1008	315	29	33
11:00	65.8	58.7	3.9	820	69.1	72.3	5.2	2.1	0.3	0.1	0.1	4.4	1007	314	29	34
Min	46.1	40.0	2.4	630	30.4	60.8	3.1	1.0	0.0	0.0	0.1	1.9	1004.0	263.0	19.0	3.0
Max	68.9	68.6	4.5	964	98.9	74.5	9.5	3.5	2.6	1.0	3.0	4.5	1010.0	336.0	35.0	64.0
Average	58.1	57.5	3.6	777	66.3	69.0	6.1	2.1	0.4	0.2	0.4	3.2	1007.5	305.2	26.5	34.8

REFERENCES

Ali, M. and M. Athar (2010). "Impact of transport and industrial emissions on the ambient air quality of Lahore City, Pakistan." Environmental monitoring and assessment 171(1): 353-363.

Beg, M. (1990). "Report on status of air pollution in Karachi, past, present and future." Pakistan Council of Scientific and Industrial Research.

Colbeck, I., Z. A. Nasir, et al. (2010). "Characteristics of indoor/outdoor particulate pollution in urban and rural residential environment of Pakistan." Indoor air 20(1): 40-51.

Colbeck, I., Z. A. Nasir, et al. (2008). "Indoor air quality at rural and urban sites in Pakistan." Water, Air, & Soil Pollution: Focus 8(1): 61-69.

EPA. "Environmental statistics, Concentration of major air pollutants (2007)." Retrieved May 28, 2010., from http://www.epa.gov.tw/en/statistics/c1020.pdf

Ghauri, B., A. Lodhi, et al. (2007). "Development of baseline (air quality) data in Pakistan." Environmental Monitoring and Assessment 127(1): 237-252.

Ghauri, B., M. Salam, et al. (1991). "Surface ozone in Karachi." Ozone Depletion: Implications for the Tropics (Ilyas M., ed.). Univ. Sci. Malaysia a nd the United Nations Environment Programme (UNEP): 169-177.

Ghauri, B., M. Salam, et al. (1994). "An assessment of air quality in Karachi, Pakistan." Environmental monitoring and assessment 32(1): 37-45.

Hashmi, D. and M. Q. Khan (2003). "Measurement of traditional air pollutants in industrial areas of Karachi, Pakistan." Jour. Chem. Soc. Pak. Vol 25(2).

Hashmi, D., G. Shaikh, et al. (2005). "Ambient air quality at Port Qasim in Karachi city." Journal of the Chemical Society of Pakistan 27(6): 575-579.

Lodhi, A., B. Ghauri, et al. (2009). "Particulate matter (PM2. 5) concentration and source apportionment in Lahore." Journal of the Brazilian Chemical Society 20(10): 1811-1820.

Oanh, N. T. K. and B. Zhang (2004). "Photochemical smog modeling for assessment of potential impacts of different management strategies on air quality of the Bangkok Metropolitan Region, Thailand." Journal of the Air & Waste Management Association 54(10): 1321-1338.

Pak-EPA (2005). "State of the Environment Report." Ministry of Environment, Government of Pakistan.

Qadir, M. A. and J. Zaidi (2006). "Characteristics of the aerosol particulates in the atmosphere in an urban environment at Faisalabad, Pakistan." Journal of radioanalytical and nuclear chemistry 267(3): 545-550.

Qadir, N. F. (2002). "Air quality in urban areas in Pakistan vs transport planning: issues and management tools." Draft paper prepared for ADB under Regional Technical Assistance 5937.

Raja, S., K. F. Biswas, et al. (2010). "Source apportionment of the atmospheric aerosol in Lahore, Pakistan." Water, Air, and Soil Pollution 208(1-4): 43-57.

Rajput, M., S. Ahmad, et al. (2005). "Determination of elemental composition of atmospheric aerosol in the urban area of Islamabad, Pakistan." Journal of Radioanalytical and Nuclear Chemistry 266(2): 343-348.

Shah, M. H. and N. Shaheen (2008). "Annual and seasonal variations of trace metals in atmospheric suspended particulate matter in Islamabad, Pakistan." Water, air, and soil pollution 190(1-4): 13-25.

Shah, M. H., N. Shaheen, et al. (2004). "Distribution of lead in relation to size of airborne particulate matter in Islamabad, Pakistan." Journal of environmental management 70(2): 95-100.

Stevenson, D., F. Dentener, et al. (2006). "Multimodel ensemble simulations of present-day and near-future tropospheric ozone." Journal of Geophysical Research: Atmospheres 111(D8).

Stone, E., J. Schauer, et al. (2010). "Chemical characterization and source apportionment of fine and coarse particulate matter in Lahore, Pakistan." Atmospheric Environment 44(8): 1062-1070.

von Schneidemesser, E., E. A. Stone, et al. (2010). "Toxic metals in the atmosphere in Lahore, Pakistan." Science of the Total Environment 408(7): 1640-1648.

Wilson, M. and J. Nicol (1994). "Tolerated noise levels in the UK and Pakistan and simultaneous thermal comfort." Renewable energy 5(5-8): 1006-1008.

World-Bank (2006b). "Pakistan: Strategic Country Environment Assessment." Volume I. Report No. 36946-PK. World Bank, Washington, DC.

World Health Organization (1984). Urban air pollution: 1973-1980, WHO, Geneva.

World Health Organization (2010). WHO guidelines for indoor air quality: selected pollutants, WHO.

Yousufzai, A., K. Hashmi, et al. (2000). "Measurements of photochemical oxidants at Sindh Industrial Trading Estate of Karachi Pakistan. ." J Chem Soc Pak 22:209–216.

Zhang, Y.-X., T. Quraishi, et al. (2008). "Daily variations in sources of carbonaceous aerosol in Lahore, Pakistan during a high pollution spring episode." Aerosol Air Qual. Res 8: 130-146.